D1291962

THE FIRST PRIVATISATION

ALSO FROM THE HISTORIANS' PRESS

THE LARGER IDEA: Lord Lothian and the Problem of National Sovereignty, 1905–1940
John Turner. 1988.

SOURCES FOR MODERN BRITISH HISTORY

REAL OLD TORY POLITICS: The Political Diaries of Robert Sanders, Lord Bayford 1910–1935
Edited by John Ramsden. 1984

THE DESTRUCTION OF LORD ROSEBERY: From the Diary of Sir Edward Hamilton, 1894–1895
Edited by David Brooks. 1987.

THE CRISIS OF BRITISH UNIONISM: The Domestic Political Papers of the Second Earl of Selborne, 1885–1922
Edited by George Boyce. 1987.

LABOUR AND THE WARTIME COALITION: From the Diary of James Chuter Ede. 1941–1945
Edited by Kevin Jefferys. 1987.

THE MODERNISATION OF CONSERVATIVE POLITICS. The Diaries and Letters of William Bridgeman 1904–1935
Edited by Philip Williamson. 1988.

THE FIRST PRIVATISATION

THE POLITICIANS, THE CITY, AND THE DENATIONALISATION OF STEEL

Kathleen Burk

British Library Cataloguing in Publication Data
Burk, Kathleen
The first privatisation: the politicians, the city and the denationalisation of steel.
1. Great Britain. Nationalised industries. Privatisation
I. Title
338.0941
ISBN 0–950890–06–5
ISBN 0–950890–07–3

PUBLISHED BY THE HISTORIANS' PRESS
9 DAISY ROAD, LONDON E18 1EA
Typeset at the University of London Computer Centre
Printed in Great Britain by Short Run Press Ltd., Exeter

. . . I am indebted to that agreeable and witty man, Lord Swinton, for the following piece of dialogue with myself:

'I believe you were the draftsman of the Bill to nationalise iron and steel.'

I admitted that I was.

'And now you are drafting the Bill to denationalise iron and steel.'

I could not deny it.

'And when our friends come back to power, I suppose you will be called upon to renationalise iron and steel.' He gazed upon me with a compassionate smile. 'Rather a Penelope-like occupation.'

This seemed to put the history of the industry into a nutshell.

Sir Harold S. Kent, *In on the Act*

TABLE OF CONTENTS

PREFACE and ACKNOWLEDGEMENTS

The reader will perhaps query the title of this book, recalling that the bill to denationalise the road transport industry received the Royal Assent on 6 May 1953, eight days before that for iron and steel. But if one makes anachronistic use of that 1980s' word, privatisation, the difference between the two becomes clear. Long distance road haulage was denationalised, but shares in the companies were not floated on the Stock Exchange for purchase by the private or institutional investor. Shares in the steel companies were, and the decisions taken and methods used had much in common with those re-invented in the 1980s.

Looking over the period from 1945 to 1988, a curious theme about the steel industry emerges. It was nationalised in 1949 by a government which did not wholly believe in its nationalisation; it was then denationalised in 1953 by a government which did not wholly believe in its denationalisation; it was then renationalised in 1967 by a government which did not wholly believe in its renationalisation. Only time will tell if it was re-denationalised in 1988 by a government which wholly believed in its re-denationalisation.

For access to manuscript collections and permission to quote from them I am indebted to the Controller of Her Majesty's Stationery Office (for material in the Public Record Office), the Trustees of the Bodleian Library, the Earl of Swinton, Churchill College, Cambridge, the Master and fellows of Trinity College, Cambridge, the British Steel Corporation, the Morgan Grenfell Group, Baring Brothers and Co. and the Bank of England. I am grateful to the following friends who read part or all of the manuscript: Robert Binyon (for Morgan Grenfell), John Orbell (for Barings), my colleagues John Ramsden and John Turner, and Jeremy Wormell, who combines historical and City skills. The responsibility any remaining errors, and for the opinions expressed, remains mine alone. Ainslee Rutledge word-processed the whole manuscript with exemplary speed and skill, and Alison Peden helped with the proofs. This book is dedicated to my daughter Miranda, and to Louise Brady, who helped make it possible.

Harwell, October 1988

INTRODUCTION

Privatisation issues have become so frequent that it is easy to forget that there had to be a first time: the denationalisation of the iron and steel industry in 1953. The philosophical and political problems involved, while difficult, were at least recognisable. Relations between government and industry during the twentieth century had gone through various phases of control and de-control, and it was largely a matter of finding the correct balance that exercised the politicians, the civil servants and the industry. Nor was it unknown for the government of the day to repeal an Act passed by a former government, although it was not then the habit that it has since threatened to become. It was when the time came actually to return the industry to the private sector that all those involved entered uncharted territory. Again, it is easy to forget that there were no precedents which the City Issuing Houses could follow in drawing up a plan to carry through the Government's policy. It was one thing to float the shares of an ordinary company on the Stock Exchange: there were rules and customs to follow in assessing the market price of a share or in drawing up a prospectus which would reveal the financial state of the company. It was quite another to do this when the Opposition was threatening to punish financially those who bought the shares, and when governmental control had interrupted the continuous financial history on which a share price should be based. In this situation, the Issuing Houses had to establish a new set of rules.

The whole episode, seen as a play, falls neatly into a Prologue, three Acts and an Epilogue. The Prologue covers the late 1920s to October 1951, but concentrates on the nationalisation of the industry by the 1945–51 Labour Government and the early contacts between the steel industry, the City and the Conservative Party. Act One covers the period from October 1951, when the Conservatives won the general election, to July 1952, when the White Paper on Iron and Steel was published. Act Two covers the passage of the Bill and the early preparations of the City, from July 1952 to May 1953. Act Three concentrates on the actual denationalisation of the United Steel Companies, the first to be sold back to the private investor, which took place in October/November 1953. The Epilogue looks very briefly at the later history of denationalisation — and renationalisation.

Of all the nationalisation exercises undertaken by the 1945-51 Government, that of iron and steel was the most contentious. It was finally passed on 24 November 1949, but in order to ease its passage through the House of Lords, the Minister of Supply, George Strauss, had pledged that no appointments would be made to the new Iron and Steel Corporation until after the next general election. On 23 February 1950 Labour won the election, but with an overall majority of only five. The general expectation was that another election would come sooner rather than later, and this the Conservatives would probably win; it therefore came as an unpleasant shock to the industry to be told by Strauss in July 1950 that the Government intended to put the Act into operation, wanted from them a list of names from which to select members for the new Iron and Steel Corporation, and planned to vest the ownership of the companies in the Corporation on 1 January 1951.

Sir Ellis Hunter was the chairman and managing director of Dorman, Long & Co., one of the major steel companies; more important in this context, he was president of the British Iron and Steel Federation (BISF). Sir Andrew Duncan had been chairman of the Executive Committee of the BISF since its founding in 1935, with a break during the war years when he was MP for the City of London and Minister of Supply. Neither was a devotee of unreconstructed free enterprise: both, in fact, believed in a large measure of governmental supervision, to which the industry had in any case been accustomed since 1932. Where they differed from Strauss was over the question of ownership. On the dubious basis that Labour had not received the support of 'a clear majority of our people', the BISF's Executive Committee refused to co-operate with the Government, declining to supply any names for the Corporation. Once the Corporation had been appointed in October 1950, however, Hunter and Duncan realised that vesting would only be stopped by another general election. Therefore they decided on a two-pronged strategy. Firstly, they would dig up any valid reason or pretext for delaying vesting which could be put to the Ministry of Supply in good faith. Secondly, in response to the Conservative Party's pledge to 'unscramble' steel if they won the next election, they would ascertain whether it could be done.

On 27 October 1950 Hunter called in at the merchant bank Baring Brothers in the City of London to talk to Sir Edward Peacock. A managing director of Barings and a former Director of the Bank of England, Peacock was probably the most influential man in the City after the Governor of the Bank of England, C.F. Cobbold. Hunter told Peacock that he, Duncan and Sir Archibald Forbes, President of the Federation of British Industry (FBI), had had a talk about the Conservative pledge to denationalise iron and steel; indeed Duncan it was who had convinced Winston Churchill, the Prime Minister, to make the parliamentary statement. But if the industry was to be returned to private ownership, Hunter, Duncan and Forbes thought it should go through the traditional channels and with the mutual consent of both sides of industry and both major political parties. By traditional channels, they meant that the securities should be sold in the usual way, and this would require the active co-operation of the City. The purpose of Hunter's visit was quite simply to begin the process of discovering whether this was possible.

On 4 December 1950 Hunter, this time accompanied by Duncan, returned to Barings, where he met with Peacock and four of his partners, and with Lord Bicester and Sir George Erskine, managing directors of Morgan Grenfell, the major 'steel House' amongst the City Issuing Houses. The lunch was somewhat disappointing for Duncan and Hunter: Peacock thought that denationalisation would be possible only if the Conservatives were quickly returned to office and Labour pledged not to renationalise. Bicester injected a note of realism, noting that it was impractical to expect Labour to make such a statement. Duncan then revealed that the Conservative Central Office had already set up a working party on steel denationalisation which wanted to contact the City, but the merchant bankers almost noticeably flinched at the thought of talking to the political working party.

What the bankers did was to constitute their own Working Committee to see what could be done. Over the next several months this Committee met in the utmost secrecy — neither the Bank of England nor the other Issuing Houses knew what was happening. By March 1951 they were able to draft a memorandum setting out in general terms how the operation might be conducted. Again, this memorandum was supposed to remain a secret; but late in March Peacock secretly loaned a copy to the Governor of the Bank of England. At the same time, a copy was given to Hunter and Duncan. The response of the Bank in its internal memoranda was strongly against denationalisation, but events would now lie fallow for some months until September

1951, when it became clear that an election was pending.

At that point events again began to move. The BISF began to be lobbied by its members about returning the industry to private hands; the Bank of England drew up memoranda setting out the problems of unscrambling; and the Treasury contacted the Bank for preliminary discussions on denationalisation, so as to be prepared for eventualities. On 25 October 1951 the general election brought the Conservatives back into power with a majority of seventeen, and in the King's Speech on 6 November, they announced that they would return the iron and steel industry to private ownership.

The period from October 1951 to July 1952, when the White Paper was published, was the most crucial of all, as the various interested parties determined what they would like to be done, what should be done, and what in fact could be done. On 20 October 1951, at the first Cabinet meeting after the Conservatives won the election, Churchill noted that the repeal of the Iron and Steel Act was one of the most urgent problems facing the new Government, and he set up a Cabinet Committee on Steel, under the chairmanship of Harry Crookshank, to decide in the first instance whether it was possible to do this before Christmas; the answer was no. However, the Government did issue an order to Stephen Hardie, the Chairman of the Government's Iron and Steel Corporation, which prevented him from making any changes in the financial or structural organisations of any of the companies.

The civil service was close behind their ministers, and on 5 November 1951 the Treasury set up a committee on steel, which included members from the Ministry of Supply, the department with responsibilities for the Iron and Steel Corporation, and observers from the Bank of England. This committee of course had no knowledge of the bankers' Working Committee, which had spent some months deciding if denationalisation was possible, and if so, how it could be done, and the Treasury committee started practically from first principles. Finally, the Ministry of Supply set up its own committee of officials.

A fortnight later the Governor of the Bank met with Peacock and Bicester, and they all decided that it would be useful to continue the discussions which had taken place in the bankers' Working Committee, this time under the auspices of the Bank of England. Their main duty would be to advise the Governor and Deputy-Governor about any recommendations the Bank might make to the Chancellor of the Exchequer, R.A. Butler. One of the first things the Governor did was to let the Chancellor know about the work the City had already put in on the problems, firstly to prevent the reinvention of the wheel, and secondly to keep a note of realism in the proceedings.

Different groups had different ends in view. The Cabinet thought that they wanted to repeal the Iron and Steel Act, but were not at all certain about what should be put in its place. Churchill himself blew hot and cold, and it seems pretty clear that behind the scenes both the industry and the Conservative backbenchers had to keep him up to the mark. Indeed, he eventually decided that road transport was an even more urgent candidate than steel for attention, and steel had to be put off from the spring to the autumn 1952 Session. The main reason for this delay, however, since Churchill would have bowed to a united Cabinet, was the Cabinet's uncertainty about what they wanted to accomplish. In short: how much control should central government retain over the activities of the companies? Those like Crookshank and Oliver Lyttelton, who preferred minimum control, were in a minority, although plenty of Conservative backbenchers shouted for this, and were effective in keeping pressure on the Cabinet. But others such as Lord Salisbury, Harold Macmillan and Butler were more moderate, and Salisbury came close to resigning over the proposed Bill. This question of control had two aspects. The first was the desire to retain, and even to extend, the powers of government to prevent undesirable developments and to compel desirable ones, and this was based on grounds of national interest as the Cabinet interpreted them. After all, there had been some measure of control since 1932, and, indeed, Hunter and Duncan of the BISF supported a government board with certain powers of overview and control. But the other aspect of control was wholly political. The central issue was, while returning ownership of the companies to the private sector, to retain enough public control to satisfy the Opposition and thereby, it was hoped, prevent the renationalisation of the industry. The City's fears here were even more short term: the bankers felt that it was vital to prevent even a threat to renationalise, or else it might prove impossible to resell the industry. Certainly the pledge on 12 November 1951 by George Strauss, now the Shadow Minister of Supply, to renationalise the industry with, effectively, no further compensation, greatly alarmed the City. This led to fruitless attempts somehow to come up with a clause in the denationalisation Bill which would guarantee the investor against such renationalisation.

In order to satisfy the Opposition, the Government attempted to draft first a White Paper, and later a Bill, which would accord with the statement made on the matter by the Trades Union Congress (TUC) in 1950. The Economic Committee of the TUC General Council, chaired by Lincoln Evans, the General Secretary of the Iron and Steel Trades Confederation, had issued a Report on Public Control of Industry. This called for a statutory board of control with additional powers to undertake schemes in the public interest, adding that public ownership need not always take the form of nationalisation. In short, the leadership of the main union in the industry did not necessarily believe in nationalisation, and certainly, although they came out publicly against denationalisation, secretly Evans told the Government to get on with it. The hope in the Cabinet, then, was that if they designed a bill which accorded with this TUC Report, Labour would not feel driven to undo it. Further, both the Conservatives and the BISF hoped that Evans would accept a seat on the planned new Iron and Steel Board, and in due course he became its Deputy Chairman.

The civil servants were all for control, especially the Ministry of Supply. They were used to it, and clearly did not trust the industry to do what was necessary in the public interest. Even in the final moments before the passage of the Bill, the Ministry of Supply tried to convince the Treasury that the bill should include provision for a levy on the industry, the proceeds of which would be directed towards desirable investment. This attempt exasperated the Bank of England, as an earlier attempt to impose desirable mergers and amalgamations before denationalisation had exasperated the Minister of Supply's Parliamentary Private Secretary, A.R.W. (Toby) Low. The conclusion of both was that the Ministry had failed to grasp what denationalisation was all about.

The Bank of England was in an ambiguous position for at least two reasons. With regard to denationalisation, the Bank's initial reaction had been one of horror at the prospect of the industry's having to go through all that again. Besides, the Bank too believed that control by government was not such a bad thing. However, once the City Issuing Houses such as Morgans and Barings were involved, the Bank accepted that what was important was a bill which would command the widest possible acceptance, whatever form it might take. Otherwise, there was grave danger that the Government's ultimate objective — resale to the private investor — would not take place. It was just as well that the Bank kept this in mind, since the Government sometimes forgot it. The other reason that the Bank was in an ambiguous position was, of course, that it was a recently nationalised industry itself. Was it the spokesman of the City to the Government, or the overseer of the City on behalf of the Government? This position was made manifestly, even brutally, clear by a partner of Rothschilds, when he accused the Bank of misusing its influence in the City, by acting as the nationalised agent of a partisan Government. Certainly the position encouraged the Bank to ensure that it had the support of the City: the Governor consulted Peacock and Bicester whenever he had an important meeting with the Chancellor scheduled.

During the period October 1951 to July 1952 the arguments and lobbying went on in public and private. After a final Cabinet crisis over the matter, the White Paper was re-drafted for the final time and issued on 28 July 1952. Duncan Sandys, Churchill's son-in-law, who as the Minister of Supply had responsibility for the Bill, had grown in authority within the Government. Not in the Cabinet, he had had to suffer the imposition of a Cabinet Committee on Steel over him. He had very definite ideas of his own about what he wanted, and was much more interested in supervision of the industry than in its return to the private sector. As to the terms of the White Paper: it proposed the establishment of two organisations. One, the Iron and Steel Holding and Realisation Agency, would assume ownership of the companies from the Government upon the passage of the new Act, and would be responsible, subject to the overview of the Treasury, for the resale of the companies to the private investor. The other organisation, the Iron and Steel Board, would be permanent, with powers to fix maximum prices and to prevent (but not compel) developments.

The debate on the White Paper on 23 October 1952 marked the beginning of the parliamentary passage of the Bill, which finally became law in May 1953. The debate, which was pretty ferocious, was not an auspicious beginning. Labour again promised to renationalise what they needed. One MP went to the heart of the matter: both sides believed in control, but the issue was ownership — who would get the profits? The Liberals came out in support of the Government. It was Strauss who pointed out one irony of the situation: while Labour under the Iron and Steel Corporation had controlled 298 concerns, the Conservatives were extending control to 2,400 firms, including many smaller companies.

Press reports were considerably more friendly than the Labour Opposition, and on 28 October 1952 the Cabinet passed the final draft of the Iron and Steel bill and authorised its introduction in the House of Commons on 5 November. The Second Reading took place on 25, 26 and 27 November. The debate was not as bad-tempered as the previous one had been. Principles were again discussed, but so were the possibilities of the Government's actually carrying though the resale. As one MP quoted from the leading article of the *Financial Times* of 26 November, Sandys 'must by now be only too acutely aware that he is proposing a Bill to sell the steel industry when there are few signs that many want to buy it.'

At the urgent behest of the Opposition, the Government set aside eight and a half days for the Committee Stage, which would be taken by a Committee of the Whole House. This lasted from 28 January to 24 February 1953 and was quite astonishingly good-tempered: while Labour made it clear that they opposed what the Government were doing, nevertheless they co-operated in order to make the final Act as workable as possible. With the decision not to obstruct, events went along at a good pace, the Opposition making a number of constructive suggestions which the Government accepted. On the final day Strauss and Sandys exchanged compliments, and indeed, this stage ended a day earlier than scheduled.

The Report stage of the Bill took place on 4, 10 and 11 March and the Third Reading on 17 March 1953. The discussion now was on problems of reselling, since many of those who might have been expected to buy back shares had expressed doubts. Strauss was worried that the Government would sell at any price just to get rid of the companies, but Sandys pledged that the Government would not sell at less than a fair price. The Bill passed by 304 to 271, a Government majority of 33, almost double their parliamentary majority. The outcome in the Lords could be taken for granted, and the Bill became the Iron and Steel Act on 14 May 1953.

Meanwhile, the Bank and the Issuing Houses had been secretly preparing for the resale for some months. Those in the City who were most involved paid the closest possible attention to what was going on down in Whitehall and especially Westminster: a Labour threat to do unspeakable things to those who bought steel denationalisation shares could threaten the whole operation.

On 16 September 1952, between the publication of and the debate on the White Paper, the Governor of the Bank, the Minister of Supply and the Chancellor decided that Sir John Morison, and accountant and senior partner in Thomson McLintock & Co., should be invited to assist the Treasury and the Ministry of Supply with the operation. No promises were made, but implicit was the assumption that Morison would in due course head the Agency which would be set up to 'own' the steel companies on their way back to the private sector. Morison prepared detailed memoranda on certain steel companies, one of which was the United Steel Companies, and in February 1953 he suggested that some of the Issuing Houses, such as Morgan Grenfell, should (together with the steel companies) draw up plans for the resale. Morgans had realised that the price set for the shares of United Steel would have to enable the Government to raise from the sale at least the amount they had paid for the shares in 1949 plus the retained profits ploughed back into the company while under state ownership. Over the next two months Morgans and United Steel, while designing a new capital structure for the company, worked to reconcile the requirements of the Government for this 'loading' with the company's financial needs.

Meanwhile, the Bank's Working Party had continued to refine their ideas, and in mid-March 1953 they reported to the Governor, who in turn reported to the Chancellor. The Governor advised that the Realisation Agency should make the offer to the public, while a Consortium of Issuing Houses would underwrite the offer. The Chancellor and the Minister of Supply agreed; but they feared that there would be political danger in the Government's appearing to approach private buyers for the steel companies' shares before the Bill had reached the statute book. However, it would be quite a different matter if the City began making preparations on its own account, and the Chancellor thought the City should get moving as quickly as possible.

The City was now to steam ahead with preparations. On 9 April 1953 the Governor met with representatives of eight Issuing Houses: besides Morgans and Barings, they included Rothschilds, Schröders, Lazards, Hambros, Helbert, Wagg and Robert Benson, Lonsdale. Morgan Grenfell, because of its position as the City's 'steel House', its work on the matter over the previous two years, and Erskine's reputation for energy and innovation, took the lead on the Issuing Houses Sub-Committee on Steel set up to work out precisely how to issue the shares. It had been hoped that the shares of six of the companies might all be sold in June and July 1953, but largely because of conditions and objections put forward by the institutions, this proved to be impossible. Only in October 1953 were the shares of the United Steel Companies offered to the public.

The flotation was initially a triumph. Because of the reluctance of the institutions, the Consortium had decided to go for the small private investor as well as for the large investor and the institutions, and prospectuses were printed in dozens of regional as well as national newspapers. Consequently, the issue was heavily oversubscribed, and applications had to be scaled down severely. But the share price fell in the aftermarket, primarily because the institutions did not make up their comparatively small allotments. The Houses had been warned: the insurance companies, for example, believed that they had only narrowly escaped nationalisation themselves under the Labour Government, and they feared that if they took too large a proportion of the shares, the Opposition might try again. Indeed, during the debates one Labour MP warned of the 'Prudentialisation' of the industry, a remark not calculated to encourage the Prudential to invest heavily.

The denationalisation of United Steel was the first of many, and the sale of six of the largest companies was completed by January 1955. In addition, a number were sold to private buyers, in some cases companies buying back subsidiaries, and by 1957, with the sale of the Steel Company of Wales, 86% of steel output was in private hands. In February 1961 most of the prior charge securities, i.e. the debentures and other fixed-interest securities still held by the Agency, were sold in one massive operation, and finally, in 1963, all of the remaining companies, except for Richard Thomas & Baldwin, were sold.

But it all went for naught, because in 1967 the Labour Government carried through the renationalisation which the industry and the City had feared in 1953. The 1967 Act followed the pattern of the 1949 Act, in that the shares in the individual companies, in this case fourteen bulk steel producers, were held by the British Steel Corporation. Two years later the Corporation was restructured and the identities of the individual companies were destroyed.

A striking point about the episode of denationalisation from 1951 through 1953 was the lack of conviction in the Government about what they were doing. Having pledged to denationalise, they were pushed into it and kept up to the mark largely by the industry and by their backbenchers. The Conservatives had, after all, accepted most of the changes Labour had made: why should the steel industry be treated differently? There was some apprehension about how the unions might react, particularly in the midst of an export drive and a threatening sterling crisis. In the crucial Cabinet meeting these points were argued, but the opposing argument was that the very principle of free enterprise was at stake. If the Government failed to denationalise, they might lose the final chance to return the people to former habits of independence and enterprise. But the amount of control imposed on the supposedly free and independent industry indicated how far in the opposite direction the Conservative Party had travelled.

A second point was a significant but unintended consequence of the denationalisation of the steel companies: wider share ownership. A primary goal of the Conservative Government from the beginning was the return of steel company shares to the hands of the former private shareholders, but this was a different matter from an attempt to increase the number of shareholders as such. The failure of the institutions to promise both to subscribe for, and especially to hold, a sufficient proportion of the steel company shares drove the Realisation Agency and the Houses to make unprecedented use of newspaper advertising, placing prospectuses in newspapers which had never before been the recipients of such attention. By the end of 1957, share ownership had spread, and steel company shares were more widely held than those of any other type of company share. But this was an unexpected outcome, rather than the result of the application of a central tenet of Conservative denationalisation philosophy.

The final point was that no one wanted responsibility for the enterprise, a sure sign of anticipated failure. The Ministry of Supply said that denationalisation would be primarily a financial problem, and managed to saddle the Treasury with the continuing responsibility, much to the Treasury's displeasure. In the City, neither the Bank of England nor the Issuing Houses wanted the responsibility for the operation itself, and in this case the Bank won, putting the responsibility on the Houses. The Houses knew their duty, and did it. It was the first such privatisation, and thus there were neither precedents to follow nor even theoretical writings to consult. The Government vaguely knew what they wanted to accomplish, but they did not know how to accomplish it. Thus the politicians turned to the City. A generation later they were to do so again, having entirely forgotten about the first time.

1 THE PROLOGUE

Of all the nationalisation decisions taken by the 1945 Labour Government, that concerning iron and steel was the most contentious. Even the Labour Government found it difficult to decide finally to nationalise the industry, since it was hard to find a non-ideological and widely-accepted argument in favour of the move: unlike the railways or the coal mines the steel industry was not loss-making; unlike the coal mines it was not suffering from the owners' unwillingness to invest; unlike the electricity industry there was no natural monopoly. Moreover, unlike all of the above, it was a manufacturing industry. Further, the leadership of the relevant unions, those in the Iron and Steel Trades Confederation, was never whole-heartedly behind nationalisation. It came down to a question of political faith: nationalising iron and steel meant entrenching Socialism in Britain. To be a good Socialist meant supporting it in order that the people, as represented by the State, would control the 'commanding heights' of economic power, enabling them both to allocate resources efficiently and to garner the profits. Thus the industry became a political football, and the whole postwar history of steel and of the processes of nationalisation and denationalisation must be seen in this context. Denationalisation could only take place once the Conservative Party were returned to power in October 1951, but the actual process began some months earlier. Indeed, the form it took is inexplicable without some knowledge of the earlier history of the industry, and of its nationalisation by the Labour Government. This chapter, therefore, will provide that background. The first section briefly sets out the relationship between Government and industry before nationalisation, the second section continues the story through the period of nationalisation itself, while the third section gives an account of the industry's relations with the City of London and the Conservative Party while Labour was still in power.

Steel and the Government Before 1945

The iron (and later steel) industry had been the backbone of the Industrial Revolution of the eighteenth and early nineteenth centuries, but by the late nineteenth century Britain was losing her position as leader amongst steel producers. By the outbreak of the First World War she ranked third behind the US and Germany as a steel producer, and by the 1920s the industry was in chronic depression. It contained too many plants and firms, equipped with inferior and out-of-date machinery, often badly located.[1] There was some limited re-organisation of the industry later in the decade, with the creation of Vickers-Armstrong in 1929 and its merger in 1929 with the steel interests of Cammell Laird to form the English Steel Corporation; this was followed by the closure of some of the works. Again, the formation of the Lancashire Steel Corporation in 1930 was accompanied by the closure of some sites and the concentration of steelmaking at others.[2] But much less rationalisation was done than was needed, the desire to remain independent predominating over arguments for efficiency. This reluctance was seldom overborne by hostile takeovers, since unwanted bids were frowned on in those more innocent days; United Steel Companies, for example, which actively rationalised and invested during the 1930s, backed off from a bid for the Lancashire Steel Corporation in 1935 as soon as the Corporation showed that it was unwelcome.[3]

The re-organisation of the industry was a concern not only of the industry itself, but of the Government and the Bank of England. During the 1920s and the early 1930s there was some discussion about about introducing protective tariffs in return for the industry's re-organising itself. But a tariff was two-sided in this context: the iron and steel industry was as likely to sit complacently behind a tariff which ensured profits for the weakest, or to restrict competition, as it was to use a temporary tariff to amalgamate and rationalise. In 1932, an Import Duties Ad-

1 K. Warren, 'Iron and Steel' in Neil K. Buxton and Derek H. Aldcroft, *British Industry Between the Wars: Instability and Industrial Development 1919–1939* (London: Scolar Press, 1979 [1982 pb edit]), p.113.

2 Warren, 'Iron and Steel', p.118.

3 Memorandum by C.F. Whigham of meeting between Benton Jones, Spens, Catto and Whigham on 1 Aug. 1935, same date, File The United Steel Companies Proposed Purchase of Lancashire Steel Corp., Box 8, Morgan Grenfell Papers (hereafter MGP), Morgan Grenfell & Co.

visory Committee (IDAC) was established, as one of the provisions of the Import Duties Act. This committee advised the Board of Trade of cases where special protection was needed, at least temporarily, in the national interest, and it decided to recommend steel. The proviso was that the period of protection would be used by the industry to re-organise, rationalise and re-equip. Leadership was to be provided by the principal trade association, the National Federation of Iron and Steel Manufacturers (NFISM).

By the time the IDAC was set up, however, the Government had given up the hope of trading tariff protection for reorganisation: rather, tariffs were an admission that free trade was permanently defunct and that for her own protection Britain needed to put up trade barriers. The IDAC could not coerce the industry, but hoped that protection would help ease re-organisation, which would depend on resolving a great many conflicts of interest. The NFISM was able to go its own way, concentrating on the development of industry-wide price associations grouped around existing Trade Associations. In short, the NFISM were primarily concerned to reconcile the wide divergence of interests within the industry, rather than to press for controversial re-structuring.

By 1933 it seemed clear that the Bank of England and the Government would be disappointed in their hopes that the industry would heal itself. Further, it was not credible to threaten the withdrawal of tariff protection, since as a consequence the industry and its workers would be even more vulnerable than before. All the Government could do was to propose, in a White Paper in Spring 1933, that the existing Trade Associations should be re-grouped and a new central body established. This was finally accepted by the NFISM in February 1934 in order to guarantee the continuation of tariffs, and with the establishment of the new central trade body, the British Iron and Steel Federation (BISF), the NFISM more or less ceased to exist. However, to satisfy the NFISM, the IDAC's original concept, of a central supervisory body which would bring about change by exerting control through a general policy of binding agreements on prices, quotas and rebates, was watered down. The new body's main task was not to re-organise the industry, but to maintain prices.

Because of disagreements within the industry, the BISF began with no members, and its survival depended on gaining the goodwill of the industry so that the various Trade Associations would affiliate. Both the IDAC and a number of steelmakers who supported the idea of a central body believed strongly that the choice of chairman would make or break the new body. He had to be acceptable both to the industry and to the Government (not to mention the Bank of England), and it took some months to find an acceptable candidate. The choice eventually fell on Andrew Duncan, chairman since 1927 of the Central Electricity Board, and a Director since 1929 of the Bank of England. His previous experience was significant: the Central Electricity Board had been developed as a 'state-backed co-ordinator of existing private undertakings', a body managed by men close to the industry.[4] The Bank of England strongly supported his appointment, as did the Government: 'forceful and ruthless', Duncan had advised the Governor of the Bank of England on the steel industry, and would continue to do so throughout the 1930s.[5] He duly became chairman of the Executive Committee of the BISF (there would be in addition a president chosen from within the industry). During the remainder of the decade the BISF continued to be primarily a price-maintaining organisation, joining the International Steel Cartel in 1935.[6]

4 Steven Tolliday, 'Tariffs and Steel, 1916–1934: The Politics of Industrial Decline', in John Turner, ed., *Businessmen and Politics: Studies of Business Activity in British Politics 1900–1945* (London: Heinemann Educational Books, 1984), pp.50–75, for the quotation as well as the basic material for this and the previous three paragraphs.

5 In 1929 the Bank set up Securities Management Trust Ltd., a wholly-owned subsidiary with only nominal capital, as the channel through which the Bank would provide funds for industrial re-organisation schemes which it supported. Duncan became Vice-Chairman of the first Board of Directors (the Governor himself was Chairman). R.S. Sayers, *The Bank of England 1891–1944*, (Cambridge: Cambridge University Press, 1976 [1986 pb edit.]), p.324, n.4, 324–25.

6 International steel cartels had existed before the First World War, but they tended to be organised around one product. Early in the 1920s, however, the most important European producers and exporters of steel joined in an attempt to establish a super-cartel to control the export of all steel products all over the world. At the end of September 1926 an agreement was signed by representatives of five national groups (from France, Belgium, Luxemburg, Germany and the Saar), setting up the first such international steel cartel, also known in Britain as the European Steel Cartel. The great slump in steel prices in 1932 re-invigorated plans to make the cartel truly international, a goal more completely achieved after 1935, with the adhesion of Britain and the US. The influence of the International Steel Cartel was greatest in 1938, but it disintegrated with the outbreak of war in

When war came the BISF became the Government's Iron and Steel Control, with a stronger hold over the industry for wartime production planning. Duncan remained in post for a short time, but in early 1940 he resigned to become President of the Board of Trade, being elected, unopposed, MP for the City of London constituency. When Winston Churchill became Prime Minister in the summer of 1940 and reconstructed the Government, Duncan became Minister of Supply, a post he held (with one short interlude back at the Board of Trade in 1941) until 1945. With the defeat of the Government in July 1945 he went into opposition, leaving politics in 1950. Of more importance for this study, in 1945 he resumed his position as chairman of the Executive Committee of the BISF, although not without some opposition from within the industry.[7]

The Nationalisation of Iron and Steel

The Labour Government's determination to take into public ownership the means of production, distribution and exchange stemmed from Clause 4 of the 1918 Constitution of the Labour Party (as amended in 1928). Whether the iron and steel industry would be part of this was a matter of doubt for some years. It was not included in the first lists of industries to be nationalised which the Labour Party drew up after the First World War; rather, it was the acute industrial distress of the 1920s culminating in the depression of the early 1930s which led to pressure for its inclusion. The Iron and Steel Trades Confederation, the trade union body,[8] submitted a scheme of re-organisation to the Board of Trade in August 1930, and in May 1931 it published a version of this proposal, which called upon the Government to place direction of the industry 'in the hands of a public (utility) corporation'.[9] Three years later the Economic Committee of the Trades Union Congress sponsored a second plan for steel nation-

September 1939. Alice Teichova, *An Economic Background to Munich* (Cambridge: Cambridge University Press, 1974), pp.138–40, 160.

7 Aubrey Jones, 'Sir Andrew Rae Duncan', in *Dictionary of Business Biography* (hereafter *DBB*), Vol. 2 (London: Butterworths, 1984), p.198.

8 The product of a major amalgamation in 1917, it was also known during renationalisation in the 1960s as BISAKTA (British Iron, Steel and Kindred Trades Association).

9 Duncan Burn, *The Steel Industry 1939–1959* (Cambridge: Cambridge University Press, 1961), pp.114–15.

alisation, which was also first devised by the Confederation, and this was endorsed by the Labour Party Conference in 1934; one concern of the Confederation was that because of inadequate co-ordination, the industry had not been able to secure what the unions considered to be good enough terms from the International Steel Cartel.[10] But in spite of the Party's endorsement of nationalisation, it did not form part of the manifesto for the 1935 election. Indeed, it nearly missed the 1945 manifesto, making a late entry only after pressure from Hugh Dalton, soon to be the Chancellor of the Exchequer, who had been Labour Party representative to the Economic Committee of the TUC, which had drawn up the 1934 plan. Pressure also came from Ellen Wilkinson, MP for Jarrow, who like Dalton, argued that the unanimous decision of the 1944 Labour Party Conference to adopt the Ian Mikardo's resolution in favour of public ownership should be heeded. The Campaign Committee of the National Executive Committee of the Labour Party agreed to its inclusion, in spite of strong resistance from Arthur Greenwood, soon to be the Lord Privy Seal, and Herbert Morrison, soon to be the Lord President of the Council and the man in charge of the nationalisation programme; both cited 'the aversion of City friends to seeing public ownership extended to manufacturing industry'.[11]

Once Labour was in power, John Wilmot, the Minister of Supply, drew up proposals for the nationalisation of the iron and steel industry, which were considered by the Cabinet on 4 April 1946. Competing with other Labour policies, such as the nationalisation of the Bank of England or of the coal mines, steel nationalisation came relatively low on the list of priorities for inclusion in the parliamentary timetable. There was little pressure for nationalisation from the Iron and Steel Trades Confederation. Labour relations within the industry were good — there had been no strikes since the General Strike in 1926 — and harmony was encouraged by the new joint consultative machinery and a shorter (48-hour) working week. Therefore, although Harry Douglass, a future general secretary of the Confederation, might tell the Prime Minister in 1947 that the union was anxious for 'full and immediate nationalisation', Lincoln Evans, who was general secretary from 1947 to 1953, opposed it, as indeed he

10 Burn, *Steel Industry*, p.116.

11 Kenneth O. Morgan, *Labour in Power 1945–1951* (Oxford: Oxford University Press, 1984), pp. 33, 95.

opposed other nationalisation schemes.[12] On the other hand, it was widely recognised that after the war the steel industry badly needed new investment and rationalisation. Therefore, in April 1946 the Cabinet reaffirmed its decision ultimately to proceed with nationalisation, but decided to defer it to the 1947–48 parliamentary Session. Meanwhile, a new Iron and Steel Control Board would be set up in consultation with the employers, to supervise development and planning and to prepare a scheme for future nationalisation.[13] Chaired by Sir Archibald Forbes, an accountant and director of Spillers Ltd who had been a Controller in the Ministry of Aircraft Production 1943–45, the Board was to remain in being from April 1946 to April 1949.

Late in April 1947, Wilmot produced a new scheme for the industry by which the Government would effectively be a holding company for compulsorily-purchased steel shares, leaving the existing companies intact; the Cabinet decided that he could introduce a bill drawn up on this basis.[14] Yet the Cabinet was very much split over the issue, and the decisive Cabinet meeting on 7 August 1947 — which took place in the midst of a ferocious sterling crisis — was long and bad-tempered. The eventual decision was that steel nationalisation would again be delayed, until the 1948–49 Session. However, this meant that it would come perilously close to the time when another election would have to be held, since the House of Lords could delay bills for two years. Therefore, the Cabinet also decided to introduce a bill in the 1948 Session to reduce the delaying powers of the Lords to one year, thus ensuring that the steel bill could become law before the next general election.[15]

The bill to nationalise the industry was introduced in October 1948 by the new Minister of Supply, George Strauss (the Prime Minister had sacked Wilmot for being too weak to stand up to the owners, or for that matter, his officials, and because the arguments over steel nationalisation had diminished his standing in

12 Morgan, *Labour in Power*, p.111. Evans wrote on 2 Oct. 1950 to Morgan Phillips, the Labour Party's general secretary, that 'Public ownership schemes of this kind were no solution to the problems of our present economic difficulties, ...' *Ibid.*

13 Morgan, *Labour in Power*, p.113.

14 Cab. 40(47), Minute 2, 28 April 1947, Cabinet Papers 128/9 (hereafter Cab.), Public Record Office, London.

15 Cab. 70(47), Minute 6, 7 Aug. 1947, Cab.128/10. Bernard Donoughue and G.W. Jones, *Herbert Morrison: Portrait of a Politician* (London: Weidenfeld and Nicolson, 1973), pp.400–403.

the Party[16]). One historian has described Strauss as 'an old Tribunite left-winger with knowledge of the scrap-metal business' who was meant to be a 'symbol of the Government's resolve in pushing on with steel nationalisation'.[17] Strauss, however, was hardly a fire-eater over steel. Hugh Dalton, who fancied himself as an unremitting radical, in 1951 referred to Strauss as 'a complete political fraud, a rich Tory pretending to be a Left Wing Socialist. He'd never wanted to nationalise iron and steel, and kept on trying to run away. We had had, more than once, to haul him back by the scruff of the neck....' [18] Nevertheless Strauss would fight to get it through Parliament. The bill was based on Wilmot's scheme with, in the end, ninety-two companies taken into public ownership; these included firms involved in the working of iron ore, the production of pig iron and steel ingots and the hot-rolling process of steelmaking. The vesting of ownership in the new Iron and Steel Corporation was planned for 1 May 1950; 'taking into public ownership' would involve the Government's exchanging gilt-edged stock, known in the City as the Steel Stock, for the issued shares, based on their Stock Exchange valuation at a certain date.

But before that could happen, the bill had to complete its parliamentary career. The House of Commons passed the nationalisation bill in 1948, but the House of Lords rejected it; it was then introduced a second time in 1949. It was again passed by the Commons, but when it went to the Lords a series of amendments was moved by the Opposition, most of which were rejected when the bill returned to the Commons. When it again returned to the Lords, they accepted the Commons' rejection of their amendments, except for two: one would make the measure inoperative until 1 October 1950 and the other would forbid the date of transfer being set before 1 July 1951. Opposition spokesmen in the Lords made it clear that they intended to stand by the principles of the amendments, their formal reason being that 'they consider that the Bill should not come into operation until the electors have had an opportunity of expressing their opinion upon it.'[19]

16 Kenneth Harris, *Attlee* (London: Weidenfeld and Nicholson, 1982), p.343.
17 Morgan, *Labour in Power*, p.117.
18 Ben Pimlott, ed., *The Political Diary of Hugh Dalton 1918–40, 1945–60* (London: Jonathan Cape, 1986), p.497.
19 469 H.C.Deb. 5s., cols 2039–2040.

In the Commons' debate on 16 November 1949, Strauss responded with the obvious argument that it was undemocratic to require two elections, especially as the Labour Party had not lost a by-election and could still claim the support of the electorate. But it was now clearly too late for the proposed vesting date of 1 May 1950 to be met without a rush, and therefore the Government would compromise. There was bound to be an election the following year, and that might mean that men who would be best suited for positions in the new Iron and Steel Corporation would be loath to throw up present employment for the uncertainty of positions in the Corporation. Therefore, the Corporation would not come into existence until after another general election. However, the bill would become law the day the Royal Assent was given, and the Corporation would be established on 1 October 1950.[20] Winston Churchill, the Leader of the Conservative Opposition, agreed to the Government's proposal and then pledged to undo the Government's dastardly work:

> Should we be returned to power one of our first steps will be to expunge from the Statute Book this wanton, wasteful and partisan Measure, in which many of those associated with it do not, in their hearts, believe, and which strikes this country a bitter blow at a bad time.[21]

On 24 November 1949 the Iron and Steel Act completed its passage through Parliament.

The general election was held on 23 February 1950, and its somewhat ambiguous result caused problems for the Government in trying to implement its plans for the iron and steel industry. Labour indeed won the highest number of seats in the House of Commons — 315 to the Conservatives' 298, the Liberals' 9 and 3 for others — but their overall majority was only 5. They had lost 78 seats compared with the 1945 results, while the Conservatives had gained 85. Worse, Labour only attracted 13.3 million votes, which, while better than the Conservatives' 12.5 million, was considerably worse than the combined votes of Conservatives and Liberals at 15.1 million. In short, while Labour's support had increased by 1.1 million since 1945, that for the Conservatives had increased by 2.5 million,[22] and this, along

20 469 H.C.Deb. 5s., cols 2041–2043. By law, a general election had to be held by July 1950.
21 469 H.C.Deb. 5s., col. 2045.
22 David Butler and Gareth Butler, *British Political Facts 1900–1985* (Lon-

with Labour's thin majority in the Commons, justified for some the view that this Government was likely to be short-lived and would be replaced by the Conservatives. This hope in turn emboldened the iron and steel industry to resist the Government's plans.

The first challenge to the Government over steel came during the Debate on the Address which began on 6 March 1950, when there was some argument over whether the Government should be permitted to implement the Act. The Commons authorised the Government to go ahead,[23] and the Minister of Supply, still George Strauss, and his officials began planning the new structure of control. Naturally, they thought it would be helpful, if not crucial, to have the active co-operation of the leaders of the industry, and accordingly Strauss invited Sir Andrew Duncan, again the chairman of the Executive Committee of the BISF, and Sir Ellis Hunter, its president since 1945, to a meeting at the Ministry on 3 July 1950.

Hunter was an accountant by training who had been managing director since 1938 and president since 1948 of the steel company Dorman, Long & Co. His time spent as a temporary civil servant in the Ministry of Munitions during the First World War had given him the opportunity of 'seeing the muddle made by the government', and left him with a somewhat jaundiced view of governmental involvement in the day-to-day workings of industry. But Hunter was no more an unreconstructed *laissez-faire* devotee than Duncan; rather, they

> shared a view of the British steel industry that was essentially new: neither laissez-faire in the British tradition nor state-organised in the German tradition.... A forceful sentence from Hunter's report of 1944 to the Ministry of Supply, might be taken as their joint text for steel policy: 'The industry, while depending on certain actions of the community, accepts the right of the community to make sure that any facilities granted are used in the public interest'.[24]

They had both strongly approved of the work of the IDAC during the 1930s, with which the BISF had closely co-operated, and they also approved of the tripartite [25] Iron and Steel Control Board which had been set up in April 1946 as a temporary

don: Macmillan Press, 1986), p.226.

23 478 H.C.Deb. 5s., cols 152–59, 1733.

24 Charles Wilson, 'Sir Ellis Hunter', *DBB*, Vol.3, pp. 390–93.

25 Made up of representatives of employers, the union and consumers.

measure. What they and the industry strongly opposed was the loss of corporate identities, private ownership and the possible loss of private managerial control.

On 3 July Duncan and Hunter went along to the Ministry of Supply. Strauss informed them that the Cabinet had decided that he should proceed with the appointment of the members of the new Iron and Steel Corporation early in October, and that the vesting date would be early in January 1951. He then asked for the BISF's co-operation in setting up the Corporation, but was told that the Executive Committee would first have to be consulted. The Committee met on the morning of 18 July, during which a politically convenient electoral theory was propounded: the Committee had given an undertaking that if the Labour Government received the votes of 'a clear majority of our people' (a quotation taken from Strauss's speech to the House of Commons on 16 November 1949), the BISF would co-operate with the Government and try to make the nationalisation scheme as least damaging to the industry as possible. But in the opinion of the Committee the Government had failed to obtain a clear majority at the election, and they therefore held themselves released from their undertaking.[26] In short, they would not co-operate by providing the Minister with a list of names from which to choose some of the members of the Corporation.

The Committee, however, had an alternative proposal for the Minister: if the Government's purpose was to ensure that the industry's policies accorded with national needs, this could be better met by means of a statutory board exercising supervisory functions. If the Minister were inclined to consider setting up an Iron and Steel Board (as distinct from a Corporation) and to postpone vesting until after the next general election, the co-operation of the industry would be promised. It was a plausible suggestion. The interim Board set up in 1946 had seemed to work well, and production had risen rapidly under its aegis. At any rate, Duncan and Hunter, at their meeting with the Minister,

26 This theory was not unique to the BISF Executive Committee. As H.G. Nicholas wrote in *The British General Election of 1950*, 'As to Iron and Steel, the Government adhered to their position that it had not properly been an election issue at all, though to the extent that it had been their majority constituted an endorsement of the action already taken. The Opposition contended that it had been a principal issue (they had frequently tried to make it so) and that the Government's narrow majority laid on them an obligation to halt the operation of the 1949 Act.' (London: Macmillan & Co. Ltd., 1950), pp.303–304.

would feel their way, and see what Strauss's reaction might be.[27]

That same afternoon, Duncan and Hunter met the Minister, and the meeting was distinctly acrimonious. Hunter opened by telling the Minister that the BISF would not co-operate by providing the requested list of names, on the grounds that the Government had failed to obtain a clear majority at the election. Strauss' response was that the BISF decision was ill-founded and ill-judged: was it the BISF's claim that a Socialist Government had to win three successive elections before carrying out the policy placed before the electorate? 'In this country we did not decide large policy issues through plebiscites but through Parliamentary elections. Elections were won not on the basis of a majority of the total votes cast but on the majority of the seats gained.' The Minister then went on to ask what the attitude of the BISF would be if he were to ask a member of the industry to serve on the Corporation, and the candidate asked for the BISF's opinion. Duncan and Hunter replied that they would not think much of the judgement of anyone who even considered such a proposal and would tell him that he would be most unwise to join a Corporation with such an uncertain future. Duncan, however, made it clear that 'there was no question of the Federation obstructing the law, still less of defying it, but in the present circumstances the Federation would do nothing more than the law required them to do in implementing the Act.' Hunter's addendum was that such a Corporation would be inadequate, and that this would be the Government's doing. Strauss responded that on the contrary, the BISF would be at fault. After this exchange of pleasantries, it was agreed that no immediate statement would be made of the position, and the Minister agreed to inform the BISF before making any announcement. The meeting ended with Strauss thanking Duncan and Hunter for coming to the meeting, adding sardonically that it had 'at least served the purpose of making the Federation's attitude abundantly clear.'[28]

On 14 September 1950 at 2pm Hunter and Duncan again met the Minister at his request: he wanted to inform them of the statement which he was scheduled to make in the House of

27 Minutes of the BISF Council, 19 Sept. 1950, Box 01546, and Notes of a meeting held in the Ministry of Supply, 18 July 1950, Box 011612, File July–Dec. 1950, both BISF Papers, British Steel Corporation Record Centre, Irthlingborough. Morgan, *Labour in Power*, pp.113–14.

28 Notes of a meeting held in the Ministry of Supply, 18 July 1950, Box 011612, File July–Dec. 1950, BISF Papers.

Commons at 2.30pm that same day. The Government intended to make certain appointments to the Corporation on 2 October 1950, and to vest as soon as possible after 1 January 1951. Once Strauss had spoken in the House, Hunter issued a statement to the press which made clear the industry's opposition to the Act and their reasons for declining to co-operate with the Government. On the same day, the Ministry of Supply sent letters to the 107 companies selected for nationalisation, suggesting that representatives of the shareholders should be appointed as soon as possible; the Ministry also requested the return by 14 October of a questionnaire on the financial structure of each company. The Council of the BISF five days later advised that these should be acknowledged but not responded to for the time being, until the BISF's Iron and Steel Act Joint Policy Committee, made up of members from the Executive Committee, could consider them.[29]

On 19 September 1950 a crucial debate on the issue took place in the House of Commons: there the Conservatives gave some indication of their strategy and restated their pledge. Churchill opened the debate by moving that the House regretted the Government's decision to bring the Steel Nationalisation Act into immediate operation during a period vital to the defence programme (a reference to the rearmament associated with the Korean War). He argued that it was not right because the Government were in a minority of 1.8 million votes[30] and because, he claimed, steel nationalisation had only been an afterthought in their manifesto. He then quoted from paragraphs 38 and 39 of the 1950 TUC Report entitled 'The Public Control of Industry', which had been drawn up by a sub-committee under the chairmanship of Lincoln Evans, the general secretary of the Iron and Steel Trades Confederation. Paragraph 38 stated that 'an alternative method of public control over private industry which deserves further consideration is the statutory Board of Control', while Paragraph 39 pointed out that such a Board might have the power to set up their own undertakings, 'either to promote

29 BISF Council Minutes, 19 Sept. 1950, Box 01546, BISF Papers.

30 A curious choice of number on Churchill's part, presumably derived by adding together the Conservative and Liberal vote (total 15,214,115) and subtracting the Labour vote of 13,266,592, to get a minority of 1,857,523. However, if Churchill had added together all of the non-Labour Party vote, he would have totalled 15,506,079, leaving Labour in a minority of 2,239,487 — surely an even better number for his rhetorical purpose? Butler and Butler, *British Political Facts*, p.226.

development which would otherwise not take place or to act as a yardstick of efficiency for the rest of the industry.' Paragraph 43, from which Churchill did not quote, ended with the statement that 'Public or common ownership need not, however, always take the form of the nationalisation of whole industries and there is important scope for selective or competitive public enterprise and for the encouragement of cooperation.'[31] What is clear is that the Conservatives from the beginning of denationalisation incorporated the ideas set out by the Confederation, partly from confidence in their worth and partly in the hope that union objections to eventual denationalisation would thereby be muted.

Churchill made it abundantly clear that denationalisation was the Conservative plan. He scorned the soon-to-be-appointed chairman of the Iron and Steel Corporation, Stephen Hardie as a millionaire Socialist[32] and pledged, on behalf of the Conservatives, that

> We shall, if we should obtain the responsibility and the power in any future which is possible to foresee, repeal the existing Iron and Steel Act, irrespective of whether the vesting date has occurred or not. We shall then proceed to revive the solution which has been set forth in the Trades Union Congress Report and which is accepted by the Iron and Steel Federation, and we shall set up again the tripartite Board, which has been proved to have worked so well. This would be the policy if we had the power, either before or after another General Election.[33]

The Federation now had the publicly-repeated pledge of the Conservative Party that the industry would in due course be returned to private ownership, and, so armed, Duncan and Hunter plotted a course of action. The new chairman of the Corporation (Hardie) wrote to the BISF on 20 October 1950, stating that the existing organisations should continue to function and that no

31 'The Public Control of Industry', Box 011612, File BISF Executive Committee, Chairman's Notes, July–Dec. 1950, BISF Papers.

32 S.J.L. Hardie was a former Chairman of British Oxygen Co. Ltd. According to the biographer of Ellis Hunter, Hardie's 'knowledge of the industries concerned was minimal and . . . [he] held his meetings down to 1951 mainly in the Dorchester Hotel.' Charles Wilson, *DBB*, Vol.3, p. 392.

33 478 H.C.Deb. 5s., cols 1719–1732. Hunter later told the directors of Barings that Duncan had convinced Churchill to make the pledge. Memo of meeting on 27 Oct. 1950, Partner's File 65A — Steel Industry 1950–51, ff.1–3, Baring Bros Archives, London.

changes would be made without full consultation. In view of this, the BISF advised their members to preserve as far as possible the independence of individual management and the common services built up by the industry over the years, such as the Research Association, the BISF itself and other related activities. It was also pointed out that because difficulties might arise over a vesting date, it was important that 'every effort should be made to find any valid reason or pretext for delaying vesting which may be put to the Ministry in good faith', although the industry was not to obstruct the Corporation in carrying out its proper duties. Indeed, the BISF and its members dragged their feet to such good effect that relations between the BISF and the Corporation at one point virtually broke down, requiring negotiations and a Memorandum in July 1951 setting out their respective responsibilities (the 'July Concordat').[34]

With defensive measures taken, Duncan and Hunter then had to ascertain whether or not the Conservatives' pledge could be carried out. It was one thing to say that the industry would be returned to private ownership; it was another actually to do it. Nationalisation had involved taking over the shares of the selected companies, while leaving the companies intact, at least for the time being; denationalisation, presumably, would mean returning these shares to their former owners. Where shares in public companies were involved, that meant the participation of the Stock Exchange, and on such matters, advice could best be given by brokers or merchant bankers in the City of London.

34 BISF Council Mintues, 7 Nov. 1950, Box 01546; Chairman's Agenda for meeting of the Executive Committee on 10 Oct. 1950, Box 011612, File July–Dec. 1950, quotation on p.3; and S.J.L. Hardie, 'Memorandum of Conference between the Iron and Steel Corporation of Great Britain and the British Iron and Steel Federation, 12 July 1951 [the July Concordat], Box 011612, File July–Dec. 1951, all BISF Papers.

The BISF And the Bankers

On 27 October 1950 Hunter called on Baring Brothers, the City merchant bank, to talk to Sir Edward Peacock. A Canadian by birth, Peacock was a managing director of Barings and a former Director of the Bank of England; he was probably second in influence in the City only to the Governor of the Bank. He had been involved with the rationalisation of the steel industry in the 1930s through links with Armstrong Whitworth, which became part of Lancashire Steel, and Barings acted for several steel companies. But it was probably less for their special knowledge of the steel industry that Hunter came to have a talk with Peacock and two of his fellow directors, Lord Ashburton and J.G. Phillimore. Rather, he came to Barings for a general assessment of the probable mood and capabilities of the City when faced with a problem for which there were no precedents and which was fraught with political traps.

Hunter informed the Barings directors immediately that he had come because of a talk with Sir Archibald Forbes and Duncan about the Conservative Party's pledge to 'unscramble' the nationalisation of steel if they won the next election. Their view was that 'any literal unscrambling whereby the former shareholders were given back their original securities would prove to be impossible', but that this would not preclude a general return to the private investor. Hunter thought that ' the more serious element in the Gov't' were uneasy at the prospect of nationalisation, but that the Labour Party had hardened in its favour in the face of the Schuman Plan [to establish the European Coal and Steel Community]. Hunter, interestingly, was confident that 'responsible opinion in the Steel Unions, personified by Lincoln Evans, disliked the project and would still be prepared to pull their weight towards an eventual set-up which, while meeting the Gov't on the question of tighter controls, would preserve the basic set-up and identities within the Industry.'

Hunter made it clear that serious thought had not been given by the industry as to what methods might be possible for a 'modified unscrambling', but that they would now be discussing it and wanted Barings to do the same. However, Hunter, Duncan and Forbes did have certain general ideas about possible procedures: they thought it should go through traditional channels, gradually and by the consent of both major parties. By traditional channels, they meant that the securities should be sold in the usual way, and this would require the active co-operation of the City. The purpose of Hunter's visit was quite simply to begin the process of discovering whether this was possible. He agreed to Peacock's suggestion that Barings consult Lord Woolton,[35] the chairman of the Conservative Party, who came to lunch at Barings on 8 November 1950.[36]

On 4 December both Hunter and Duncan returned to Barings for lunch, where they again met Peacock and Ashburton and two other directors.[37] But this time two others joined the meeting, Lord Bicester and Sir George Erskine of Morgan Grenfell, another leading City merchant bank and Issuing House. Morgans were brought in for several reasons. First of all, Bicester, head of his House and a current Director of the Bank of England, combined great City prestige with political links, having been for many years chairman of the City's Conservative and Unionist Association. Secondly, Morgans were the City's 'steel house', having on their list of clients six steel companies, including five of the six largest. Thirdly, stemming from these connections, Bicester and Erskine themselves had great expertise in steel, having worked on a number of rationalisation schemes and issues of shares since the early 1930s. And, fourthly, Erskine was legendary in the City for his energy and his innovative mind.[38]

Peacock observed that he had gathered from the talk with Hunter that time was of the essence: if denationalisation did not come within a month or two of vesting day (15 February 1951), it would rapidly become less and less feasible. For one thing, the individual companies might be amalgamated into new companies and would therefore have no loyalty on which to call nor earnings record on which to be judged; for another, ' the big institutional holders were [no longer] likely to have retained their Compensation Stock and so [no longer] be in a position to hand it back again in exchange for their previous holdings of steel company securities.' Peacock also emphasised that in his

35 Notes of meeting, 27 Oct. 1950, Partner's File 65A — Steel Industry 1950-51, ff.1–3, Barings Archives.

36 Ashburton to Woolton, 27 Oct. 1950, f.5, 'We have got one thing we should much like to discuss with you in regard to the Steel Industry.' and Woolton to Ashburton, 31 Oct. 1950, f.6 'I shall be very glad of the opportunity of a talk with you and your colleagues.', both Partner's File 65A, Barings Archives.

37 Evelyn Baring and C.H.G. Millis.

38 Kathleen Burk, *Morgan Grenfell 1838–1988: The Biography of a Merchant Bank* (Oxford: Oxford University Press, forthcoming), Chapter 6.

opinion, steel could only be resold to the public if there was bi-partisan agreement not to revive the question of nationalisation. Finally, the process of resale would have to be done in stages and would take several months to complete.

Duncan stated quite frankly that he found these opinions disappointing. For one thing, it would be 'impossible for the Socialist Party to promise' not to renationalise. Further, he would not be able to keep all of his colleagues in the steel industry in line unless he could hold out to them a definite promise that a Conservative Government would promptly denationalise; indeed, there would already have been defections if 'he had not, on his own responsibility, assured Mr Churchill that "unscrambling" was feasible and got him to make his recent pronouncement in Parliament to this effect.' In short, Duncan considered denationalisation essential, and 'he looked to the City to find the way to do it'.

Hunter pointed out that the financial position of the companies was strong and, he claimed, the industry's efficiency was equal to the most efficient anywhere. At this point Bicester joined the discussion. He recognised that it was impractical to expect the Socialist Party to promise to leave steel in private hands, but in that case, the companies' shares had to be attractive enough to encourage the investor to take the chance. In these circumstances, 'the public would probably only be induced to invest in Steel again if the companies distributed rather larger dividends than at present'.[39]

Duncan then revealed that the Conservative Party's Central Office had set up a working party to prepare a plan for steel unscrambling, and that R.A. Butler, the Chairman of the Conservative Research Department, was taking a special interest in it. The working party was anxious to be put in touch with the City to discuss the financial aspects of the plan. The bankers, however, almost noticeably flinched at this: as Peacock said, they were prepared to do all they could to help, but it was essential that Barings and Morgans consider the matter further amongst themselves before they received a visit from the political working

39 The companies had maintained a low dividend policy after the war, preferring to plough profits back into modernisation of plant rather than distribute them to shareholders; as a result some steel shares had been trading at a discount to their asset value on the Stock Exchange on the date set by the Act and the resulting takeover or nationalisation price levels had been considered confiscatory by many.

party.[40]

What the bankers did was to constitute their own Working Committee to see what could be done. On 19 December 1950 Peacock, Ashburton and Phillimore of Barings again received Bicester and Erskine of Morgans, but this time they were joined by Sir Archibald Forbes, chairman of the FBI and formerly chairman of the Labour Government's Iron and Steel Control Board 1946–49, Sir John Morison, senior partner of the accountants Thomson McLintock, who would eventually be the first chairman of the Conservatives' Iron and Steel Holding and Realisation Agency, and T.S. Overy, founder and senior partner of the solicitors Allen and Overy. Peacock had clearly become more enthusiastic since the previous meeting, announcing at the outset that the denationalisation of steel was important, although adding that the City had never handled anything so big. He also felt that it was impractical to return steel securities to the same hands, even if it was desirable: the problem should be considered as one of selling the nationalised industry to the investing public.[41] This was indeed the eventual decision, but not before considerable heart-searching over the problem in the City and the Government.

Overy set the scale of the problem: in addition to the gilt-edged stock shortly to be issued as compensation for the securities vested (i.e., the shares in the companies being nationalised), the Iron and Steel Corporation would soon be borrowing in order to provide itself with working capital. Therefore, the total figure with which the City would have to deal was likely to be more than £300 million. At this point, Ashburton sounded a note of caution. Resale to the public would be very difficult in any case if Labour pledged to renationalise the industry at the same takeover values, but he 'also thought that insurance companies such as the Prudential, who themselves had so narrowly escaped nationalisation, might be chary of antagonising the Socialists by collaborating in the de-nationalisation of Steel.'[42] There was no

40 Memorandum of meeting, 5 Dec. 1950, ff.8–11, Partner's File 65A, Barings Achives. The relevant file in the Conservative Research Department Papers is silent on the subject of working parties on steel denationalisation, whether their own or the City's. File CRD 2/3:1, Conservative Party Archives, Bodleian Library, Oxford.

41 Minutes of Working Committee, 19 Dec. 1950, Partner's File 65A, f.13 (but 4pp long), Barings Archives.

42 *Ibid*.

further discussion on this point, which would prove to be crucial when the shares in the first company were offered to the public.

Phillimore suggested that perhaps the Iron and Steel Corporation should remain in being to hold the fixed-interest debt (representing loans to the company which required regular interest payments, such as debentures and loan stocks, rather than the ordinary shares or equity that represented actual ownership), to provide new finance for the industry out of the proceeds of ordinary shares sold to the public and to exercise the function of control as formerly exercised by the Iron and Steel Control Board. Overy, however, insisted that the industry would oppose the Corporation's remaining in being. Forbes then suggested that perhaps the first move should be for the Corporation to transfer all of the steel securities to a huge holding company.[43] Phillimore's and Forbes' suggestions contained the germ of the future structure of denationalisation and control, in their outlines of an Iron and Steel Board and a Realisation Agency.

It remained only to discuss the relationship between the future (very secret) City discussions and the politicians. Ashburton for one thought it would be a mistake for the City to become involved in the political aspects by meeting with the Conservatives' working party, as Duncan had requested on 5 December. It would be far better for the bankers to confine themselves to dealing with the industry, and to let the industry itself keep contact with the Conservative Party if they wished to do so. Meanwhile, the City would get to work.[44]

On 28 December Forbes, Erskine, Morison, Overy, Ashburton and J. Ivan Spens, an accountant who was on the board of The United Steel Companies, and who had been invited to become a member of the bankers' Working Committee at Bicester's request, met at Barings to begin examining the problems in greater detail. Over the next several months they met in the utmost secrecy — neither the Bank of England nor the other Issuing Houses[45] knew what was happening. Certainly no rumours seem to have leaked to the press — or if they did, nothing appeared in print. Each member of the Working Committee received one copy of the Minutes, which were numbered. Forbes was the chairman, and his intimate knowledge of the steel industry made him the

43 *Ibid*.
44 *Ibid*.
45 Firms which sponsored the issue and sale of securities to the investing public.

natural liaison with the BISF, and specifically with Duncan and Hunter. At the meeting on 28 December, Ashburton had commented that it would now be useful to have some guidance as to the political snags, and it was agreed that Forbes would also liaise with the Conservative Central Office through Duncan,[46] who in addition maintained close contact with Churchill. Therefore, in the industry only Duncan and Hunter knew about the City discussions; in the political world it is unclear who knew besides Butler and possibly Churchill, but presumably whoever made up the Conservative Party's working party had some inkling that talks were going on, even if they remained in ignorance about who was doing the talking; and in the City, only the members of the bankers' Working Committee, and the senior managing directors of their banks, knew about them.

While discussions were continuing, the vesting date (15 February 1951) drew closer, and the Conservatives made one last attempt in the House of Commons to stop the process. But the Government, which had only a majority of five in the House, won the division by ten votes, leading Churchill to mutter to his colleague Oliver Lyttelton that 'It looks as though these bastards can stay in as long as they like.'[47] The Liberals thereafter made it clear that they would not join the Conservatives in any further attempt in the current Parliament to reverse the steel decision, and, as *The Economist* pointed out, this emphasised what the Tories now recognised: they needed a general election in order to denationalise steel.[48]

By 8 March 1951 the bankers' Working Committee were ready to draft a report. Erskine at this point thought that the equity of all of the major companies could be dealt with in one operation, provided that they could get the City sufficiently interested in the operation. Each House could take half of its participation for long-term investment and underwrite the other half for sale on to investors. This was an idea which would continue to attract, but would in the end be seen to be impractical. Beyond this, the Working Committee agreed on certain principles which should govern the transaction, a situation for which there were absolutely no precedents: (1) the sale should be made to the public

46 Working Committee, 2nd meeting, 28 Dec. 1950, Partner's File 65A, f.17, Barings Archives.

47 483 H.C.Deb. 5s., 1742-1871. Quoted in Pimlott, ed., *Diary of Hugh Dalton*, p. 499.

48 *The Economist*, 10 Feb. 1951.

and not merely to the former shareholders; (2) the price should be the price judged correct in market terms for each company at the time of the sale, not the price of the original takeover; (3) a fair price should be established by taking into account dividend policy and historic and expected earnings; (4) on this basis, pink forms (for preferential allocation in case of over-subscription) might be issued to shareholders on the register on the vesting day (15 February 1951); and (5) Steel Stock (the gilt-edged stock issued by the Government as compensation for the steel securities taken over), as well as cash, should be accepted in payment for any allotment made, to avoid shareholders having to sell this or other stock in the market to raise money. But these technical points were irrelevant unless the Conservatives came back with a good working majority at the next election and remained in office for four or five years. Forbes and Morison then set to work to draft a report for Duncan.[49]

On 19 March the Working Committee met at Barings to consider the draft Report. Forbes and Morison had made one major change: because it would be necessary to carry through denationalisation as speedily as possible, and in order to retain the goodwill of the industry, they had decided that it was impracticable to sell at the market price; selling at the takeover price would be the only workable basis. (The reason, presumably, was that some of the less profitable firms would not command a very high market price, and their shareholders had done better selling to the Government than if they had retained their interest in the firms.) This was eventually accepted, although Phillimore of Barings was not fully convinced. The Report was then taken away for discussions with senior directors, subsequently amended and amplified and then reissued in its final form.[50]

The Report had been commissioned by Hunter and Duncan of the BISF, and the Committee was anxious to receive their reaction. Forbes met with them, and was able to report that they had received it 'very well indeed'. Duncan, however, felt very strongly that the time scale was finite: they could hold the position for eighteen months, but he had grave doubts about any longer period.[51] There was also the question of loose talk — they

49 Working Committee meeting, 8 March 1951, f.40, Partner's File 65A, Barings Archives.
50 Notes of a meeting, 19 March 1951, Partner's File 65A, f.41, Barings Archives.
51 Notes of a meeting, 6 April 1951, Partner's File 65A, f.49, Barings

did not trust the politicians. Peacock, Bicester and Forbes had already agreed on the draft of a letter which briefly set out the background to the Report, but emphasised that its conclusions were very tentative and the document should not be passed on. The letter stated that if conditions did not deteriorate, it should be possible to carry out the operation, but it was vitally necessary that the statement of terms on which it was to be done should be kept very general. In short, ' nothing should be said other than that it was the intent of the Conservative Party to restore the steel industry to private ownership in a way to give the former owners the chance to reinvest in the Trade.'[52] Armed with this letter, Duncan had seen Churchill, 'who had given an absolute promise that no politician attached to the Conservative Party would make any statement beyond what was suggested in the Peacock/Bicester letter and would in fact do nothing until [Churchill] had been told by Duncan that things had been completely worked out and they could say something about it.'[53]

By early April 1951, then, the leaders of the industry and of the Conservative Party had both been apprised that denationalisation was a feasible proposition, but both sides had been sworn to secrecy about the proposed methods. Revelation should have ended there, but Peacock decided that it was imperative that one further institution be told: the Bank of England. It is impossible to know whether he consulted anyone about this, since apparently no written reference exists which might give a clue. He might have consulted Bicester; but if he did, it is strange that Bicester, a current Director of the Bank, did not himself give a copy of the Report to the Governor, C.F. Cobbold. He did not: but Peacock did, and as soon as possible.

The Bank made a copy of the Report, which was handed back personally to Peacock on 2 April, and then proceeded to analyse it.[54] In sum, they were nowhere near as sanguine as the merchant bankers. As one official wrote,

> It seems to us that the prospect of the industry becoming a shuttlecock in the political game must militate against the re-issue of

Archives.

52 Peacock and Bicester to Forbes, 29 March 1951, Partner's File 65A, f.44, Barings Archives.

53 Notes of a meeting, 6 April 1951, Partner's File 65A, f.49, Barings Archives.

54 'Iron and Steel Industry', 21 March 1951, plus comments thereon, File G1/125, Bank of England Archives.

> the Companies' Securities to the public without considerable loss. Moreover there are so many problems and difficulties involved in any form of unscrambling that despite their often reiterated pledge to repeal the Act the Opposition would be well advised to confine themselves in the future to a more guarded promise to consider whether and to what extent it would be practicable....
>
> If, however, on coming to power the Opposition decide upon "unscrambling", it would in our view be essential that such an operation should be conceived in the simplest possible terms, namely a straightforward offer of the shares for subscription by the public for cash. That in itself would be so huge an operation as to tax the Companies' (and Issuing Houses') resources....[55]

But unless and until the Opposition came to power, there was nothing to be done by anyone, and the matter was put aside until September 1951. Then events began to move. On 19 September the Prime Minister announced to the Cabinet that he intended to call an election,[56] which was set for 25 October. The following day the Governor of the Bank, on the assumption that denationalisation would be part of the Conservatives' election manifesto, asked that the Chief Cashier, P.S. Beale, draft a scheme of denationalisation, along with a note on the objections to such a scheme and any possible alternative lines of action.[57] Beale produced his outline and assessment a week later, and his conclusion was uncompromising: 'Nothing seems to alter the conclusion that unscrambling in the sense of attempting to re-market the shares is not a practical proposition. It would still seem more fruitful to approach the problem with the object of securing decentralisation of control and management — not shareholding.'[58]

That may have been the rational conclusion, but it was not the political one. The Conservatives did indeed include a pledge in their manifesto:

> The Iron and Steel Act will be repealed and the Steel Industry allowed to resume its achievements of the war and post-war years. To supervise prices and development we shall revive, if necessary with added powers, the former Iron and Steel Board representing

55 Memorandum 'Unscrambling', 27 April 1951, File G1/125, Bank of England Archives.
56 Pimlott, ed., *Diary of Hugh Dalton*, p.555.
57 Peppiatt to Chief Cashier, 20 Sept. 1951, File G1/125, Bank of England Archives.
58 P.S. Beale, 'Unscrambling of Steel', 28 Sept. 1951, File G1/125, Bank of England Archives.

the State, the management, labour, and consumers.[59]

However, this pledge was as far as any Conservative speaker would go during the campaign: Forbes had double-checked with Duncan, who had assured him that 'no Conservative speaker in the election would indulge in embarrassing statements', but would keep to the Peacock/Bicester letter.[60]

The bankers' Working Committee reassembled in view of the pending election, and Forbes reported to it the gist of his recent conversations with Hunter and Duncan. First of all the 'troops were in good heart' and although Hardie of the Corporation planned certain amalgamations (in South Wales), he had not yet done anything. He was so maladroit, Forbes added, that he had united the ranks of the steel industry against him. The Committee again decided to eschew any direct political liaison with the Conservative Party until after the election, but set in motion work by T.S. Overy, the solicitor, on sections of the Iron and Steel Act which would have direct relevance on the operation.[61] This information would, in due course, be conveyed to the new Government.

Within the industry itself, members of the BISF began to lobby their leaders about returning the industry to their hands, without, however, knowing about the plans already made; certainly the companies themselves had done nothing.[62] Duncan went some way towards reassuring them at the meeting of the Executive Committee of the BISF on 16 October, the week before the election: in the first mention to the industry of the City discussions, he reported that 'it is understood that the financial aspects of "unscrambling" have already been thought out in some detail

59 'The Manifesto of the Conservative and Unionist Party — General Election, 1951', p.5, Box 011612, BISF Papers.

60 Working Committee, 27 Sept. 1951, Partner's File 65A, f.72, Barings Archives.

61 Working Committee, 27 Sept. 1951, f.72 and T. Stuart Overy, 'Memorandum', 4 Oct. 1951, f.78, both Partner's File 65A, Barings Archives.

62 'It is not known whether or not the Conservative Party has any plan for this restoration and so far as is known the management of the confiscated iron and steel companies, now called publicly owned companies, have made no plan nor have they begun to examine the problems of restoration.' W[alter] B[enton] J[ones], 'Return of Iron and Steel Industry to Free Enterprise', 3 Oct. 1951, Box 011612, File July–Dec. 1951, BISF Papers. Benton Jones was chairman of United Steel Companies Ltd.

in the City and that no insuperable difficulties are involved.' But he also felt it wise to remind the Committee, and through them the industry, that

> apart from the restoration to private ownership of the companies now publicly-owned, the ultimate objective must be to secure a solution sufficiently broad to remove the industry from the political arena for all time.... any alternative proposals should be so framed as to ensure... the support of the Steel Unions... the Liberal Party and, indeed, the less extreme elements of the Labour Party.
>
> There are grounds for believing that Mr. Lincoln Evans would not oppose, and might even support, a measure restoring the Industry to private ownership, provided that this were coupled with definite provisions for effective statutory control of the Industry.[63]

There were implications within this statement which parts of the industry would find most unpalatable, and it would in due course require no little effort on Duncan's and Hunter's part to reconcile them to what was necessary.

And so, with an election campaign in full swing, both the industry and their City contacts were as ready as they could be before the result was known. By October the outcome seemed pretty clear — Dalton wrote in his diary on 13 October that 'I think we're out'[64] — and this brought a new set of players into the game: the Civil Service. Sir Edward Bridges, the Permanent Secretary to the Treasury, sent word to Cobbold, the Governor, that he would like a preliminary word with the Bank about steel denationalisation 'so as to be prepared for eventualities. (Nothing to be put in writing.)'[65] By 12 October Sir Wilfrid Eady, Joint Second Permanent Secretary to the Treasury, was already making plans to rebuild the industry into about sixteen units through amalgamations and rationalisation before they were returned to private ownership.[66] Certainly many civil servants were reluctant to disengage from the industry, and after the election both the industry and the Bank were to face difficulties in curtailing their interventionist enthusiasms.

On Thursday, 25 October 1951, the Conservative Party won the general election. If the results in 1950 had been ambiguous, those

63 Chairman's Agenda, Executive Committee, 16 Oct. 1951, Box 011612, BISF Papers.
64 Pimlott, ed., *Diary of Hugh Dalton*, p.560.
65 Note by Cobbold, 5 Oct. 1951, File G1/125, Bank of England Archives.
66 H.C.B. Mynors, 'Steel', 12 Oct. 1951, File G1/125, Bank of England Archives.

in 1951 were even more so: although the Conservatives won 26 more seats than Labour, and had an overall majority of 17, they polled 231,067 fewer votes. This result undoubtedly contributed to the caution with which they approached the repeal of Labour measures; and if one theme dominated the subsequent period of denationalisation, it was the Conservatives' lack of a united, firm belief in what they were doing.

2 CONSTRUCTING THE WHITE PAPER

The period from the Conservative victory on 25 October 1951 to the publication of the White Paper on 28 July 1952 was one of uncertainty and confusion. From the political point of view, it was the most crucial period of the whole denationalisation episode, as the various interested parties determined first what they would like to be done, then what probably should be done, and finally what in the end could be done. For the industry, and in particular for the BISF, there were two immediate tasks: they had to make certain that the Conservative Government actually included steel denationalisation in the King's Speech and thereafter found time for it in the parliamentary timetable; and they had to ensure that Hardie and the Corporation were prevented from taking any irrevocable decisions about the industry. The Cabinet for its part had to decide upon the priority to be given to steel, and then how to re-organise the industry's relationship with Government. The civil service, even before the King's Speech, had begun setting up committees to deal with steel denationalisation. And the City, with the Bank of England now assuming leadership, continued to try to work out how actually to resell the industry, as well as to guide the politicians in taking decisions which would facilitate — and not hinder — the task.

The main theme of this chapter, then, is the transformation of a political promise into a legislative measure. The first section shows how the political parties established their positions. The second section describes how Whitehall and the City co-operated in drafting a Bill, trying to accommodate the requirements of both the industry and the Cabinet. The reaction to the draft measure was mixed, and the third section concentrates on the resultant Cabinet crisis, as Ministers argued over the Bill — Lord Salisbury threatening resignation — and finally resolved their differences over both strategy and tactics.

The Opening Exchange of Shots

Once the Conservatives had won the election, Duncan again swung into action. He had apparently brought about Churchill's parliamentary pledge to denationalise steel, and its inclusion in the manifesto: now he had to ensure that it became part of the King's Speech. This was not a foregone conclusion, since only a few of the many manifesto commitments could be dealt with in the approaching Session. But by the end of October Duncan was helping to draft the relevant part of the King's Speech, which was scheduled for 6 November. At the same time, the BISF officers, with the help of their legal advisers, were (at the Prime Minister's request) drawing up a memorandum setting out lines on which a short Bill might be drafted, which would suspend the powers of the Corporation and set in motion the denationalisation of the industry.[1]

During the first Cabinet of the new Government, held on 30 November 1951, the Prime Minister urged that they consider ways to repeal the Iron and Steel Act. If a short and simple Bill could suffice, it might be possible to pass it before Christmas; if, however, the necessary legislation was complicated, then it would have to wait. A Committee under the Minister of Health, Harry Crookshank, was appointed to consider which course to take and report to the Cabinet; one member was the Minister of Supply, Duncan Sandys (not in the Cabinet), and at his request, Sandys sent him the BISF's memorandum on a possible Bill. The Committee reported back to the Cabinet on 1 November and again on 2 November, and their conclusion was that there should be no legislation before Christmas. On the whole, the BISF agreed with the decision, since 'the delay will give ample time for drafting a comprehensive Bill'.[2]

The BISF also very much agreed with another decision taken by the Cabinet, acting on the advice proffered the month before

1 Working Committee, 29 Oct. 1951, Partner's File 65A, f.86, Barings Archives. Chairman's Agenda for Executive Committee, 20 Nov. 1951, Box 011612, File July–Dec. 1951, BISF Papers. The Legal Advisers were Sir Geoffrey Cox, the Parliamentary Agent, and Walter Gumbel, retained as Counsel.

2 Cab. 1(51), Minute 4, 30 Oct. 1951; Cab. 2(51), Minute 1, 1 Nov. 1951 and Cab. 3(51), Minute 2, 2 Nov. 1951, all Cab. 128/23. The other members of the Cabinet Committee were the Chancellor of the Duchy of Lancaster, Lord Swinton, and the Attorney-General, Sir Lionel Heald. Chairman's Agenda, Executive Committee, 20 Nov. 1951, Box 011612, File July–Dec. 1951, BISF Papers, for the quotation.

by T.S. Overy for the bankers' Working Committee, the BISF itself, and then the new Government: that it would be perfectly legal for the Government merely to issue a directive to the Corporation (rather than legislate) to refrain from doing anything which would prejudice the position until legislation could be passed. The BISF were anxious because on 16 October, the Chairman of the Corporation had detailed his plans for the regional grouping of companies, irrespective of products or other technical considerations, with the beginning to be made in South Wales. They were afraid that he would break up and reassemble companies in a way that would make their denationalisation difficult, if not impossible: Duncan only found out later that 'before the change of Government, the permanent officials at the Ministry of Supply were going to recommend to the late Minister that Mr Hardie's plan was unsound and that he should be prevented from going ahead with it.'[3] On 12 November, then, during the Debate on the Address, the new Minister of Supply announced that in order to hold the position pending the passing of legislation, the Corporation could not, without his consent, alter the financial structure or management of any company under its control or sell or dispose of any part of any such company. The directive was issued on 13 November, and duly sent by Hardie to the chairmen of all the relevant companies.[4]

But meanwhile, on 6 November 1951, came the King's Speech to Parliament and the moment for which the BISF had long been waiting: 'A bill will be placed before you to annul the Iron and Steel Act with a view to the reorganisation of the industry under free enterprise but with an adequate measure of public supervision.'[5] On 12 November Sandys gave some indication of the ideas which the Government were developing:

> We, on this side, have always recognised that there must be public supervision, particularly in the field of development policy and prices. We have never, on the other hand, accepted that adequate public supervision can only be obtained at the price of nationalisation with all its restricting effects on the industry. We are supported in this by the much quoted report of the T.U.C. last year,

3 Cab. 2(51), Minute 1, 1 Nov. 1951, Cab. 128/23. T.S. Overy, 'Memorandum', 4 Oct. 1951, Partner's File 65A, f.78, Barings Archives. Chairman's Agenda, Executive Committee, 20 Nov. 1951, Box 011612, File July–Dec. 1951, BISF Papers, for the quotation.

4 Executive Committee, 20 Nov. 1951, Box 011612, BISF Papers.

5 493 H.C.Deb. 5s., col.52.

> in which the view is clearly expressed that public ownership is not the only practical form of securing adequate public control and that a board, possibly on the lines of the former Iron and Steel Board, if necessary with additional powers, offers an effective alternative to nationalisation....
>
> The creation of such a board, representative of Government, management, labour and consumers, will, of course, be an essential feature of the proposals which we shall, in due course, submit to the House. Such a board, unlike the Corporation, could embrace the whole industry, which would be an additional advantage.[6]

The Conservatives evidently hoped that Labour would be reconciled to the repeal of the Iron and Steel Act by the promise to maintain public supervision of the industry. But it was not to be. Strauss answered with outright defiance:

> ...we propose to resist at every stage and oppose with all the powers at our command any proposal to remove from the authority of Parliament the broad, effective and positive control which it now possesses over the iron and steel industry of this country. If the present Government, nevertheless, succeed in removing that control, we will seek to restore it at the first opportunity.
>
> Moreover, we say that, if all or any section of this industry which is now to be handed over to private investors, when the time comes, as it inevitably will, to re-transfer the shareholdings to public ownership, those investors should be compensated on the principle that under no circumstances will the total compensation already paid out be increased.[7]

The speeches of the Minister and Shadow Minister of Supply excited a range of reactions. *The Economist* wrote that 'Few debates in recent years have revealed more bitterness of feeling between the parties and none has helped to create a political situation more full of difficulties and dangers.... The party leaders are indulging, and many must know they are indulging, the worst passions and prejudices of their supporters.'[8] This was one topic on which the Labour Party could unite, and their anger was given added heat by the fact that it was not the habit for governments to repeal immediately the Bills passed by a previous government. The anger felt by Labour over the Conservatives' plans for steel would find a focus in the debate on the White Paper the following October.

6 493 H.C. Deb. 5s., cols.675–76.
7 493 H.C. Deb. 5s., cols.665–66.
8 *The Economist*, 17 Nov. 1951, pp.1166–67.

While the officers of the BISF were reasonably satisfied with Sandys' speech, there was dissatisfaction from a large number of predominantly smaller firms who had not been nationalised by Labour. The Conservatives proposed that these firms, previously subject only to general ministerial control on supplies and prices, should now come within the control of the proposed new Board, which was to be given powers in relation to development as well as supplies and prices. This dissatisfaction made itself known at the first Council Meeting of the BISF which took place after the debates (on 18 December 1951). In reply, Duncan and Hunter

> pointed out that the proposals now made went no further than, and in some respects did not go so far as, had been agreed at the time of the discussions in 1947 as the Industry's alternative to nationalisation.... The Industry and the Government were committed to the formation of a statutory rather than an advisory Board and, indeed, such a Board, with comprehensive powers over the whole Industry, was the only basis on which it was likely to be possible to counter the Opposition threat of re-nationalisation.[9]

The question of whether the Board should be advisory or statutory was one on which the Government had asked the BISF's advice: an advisory board would leave the power in the hands of the Minister, while a statutory board would have its own powers and only be responsible to the Minister, and so to Parliament, in a very general way. (The 1946–49 Board had not been statutory.) The thrust of the industry was always to keep itself as far away from direct political interference as possible, and in the circumstances they had recommended a statutory board. This was all explained to the Council members, and after a 'full and frank discussion', they unanimously endorsed the recommendations which Hunter and Duncan had made to the Minister of Supply.[10]

Another reaction, that of grim satisfaction at the exchange, was expressed in his diary by the Labour backbench MP, Richard Crossman: 'Actually, the last word had been said in the first speech, by George Strauss, when he announced that, if returned to power, Labour would reverse the denationalisation of steel and would make sure that no further compensation was paid as a result of the Tory denationalisation. This will make denation-

9 BISF Council Minutes, 18 Dec. 1951, Box 01546, BISF Papers.
10 *Ibid.*, for the quotation. Chairman's Agenda, Executive Committee, 20 Nov. 1951, Box 011612, File July–Dec. 1951, BISF Papers.

alisation extremely difficult, since who is going to provide the money in a free market, under such risk?'[11]

Indeed, if Crossman had been aware of just how much consternation was caused in the City by Strauss's speech, he would have been delighted. The bankers' Working Committee, meeting the following day, were very worried by it, in particular by his threat that under no circumstances would compensation be increased. When the companies had been taken into public ownership, shareholders had been given British Iron and Steel Corporation Stock paying 3.5% interest; Strauss had now threatened that if, during the period between denationalisation and Labour's promised renationalisation, dividends were paid to shareholders of the companies in excess of 3.5% per year, full account would be taken of these gains in any future compensation. The Committee believed that this would have a 'very harmful influence on a prospective steel investor'. It was decided that Peacock, Bicester and Forbes should discuss the threat with the Governor of the Bank,[12] who conveyed the fears to the Government;[13] and some weeks were lost in fruitless attempts to come up with a clause in the denationalisation Bill which would guarantee the investor against loss.

11 Janet Morgan, ed., *The Backbench Diaries of Richard Crossman* (London: Hamish Hamilton and Jonathan Cape, 1981), p.34.

12 Working Committee, 13 Nov. 1951, Partner's File 65A, f.98, Barings Archives.

13 'It is felt by those who have spoken to me that if the general climate were favourable, the many technical difficulties of a refinancing operation could be faced and in some way or other overcome; but if that it were to be the focus of prolonged political controversy on the lines of the recent Debate in Parliament the prospects for such an operation would indeed be black.' Governor to Chancellor, 17 Nov. 1951, File G1/125, Bank of England Archives.

The Government, the City and the Drafting of the Bill

Meanwhile, the relevant departments of Government were organising themselves to gather information, consult with interested parties,[14] draft a Bill and get it through the Cabinet. First off the mark was the Treasury. Sir Edward Bridges, the Permanent Secretary to the Treasury, had already contacted the Governor on 5 October to ask for information 'to prepare for eventualities', and on 29 October the Governor sent him a memorandum on some of the financial points. On the following day the Chief Cashier of the Bank, P.S. Beale, with two other Bank officials attended a meeting at the Treasury with Sir Bernard Gilbert, the Joint Second Permanent Secretary to the Treasury, Sir Archibald Rowlands, the Permanent Secretary to the Ministry of Supply, and Eady. At this meeting Rowlands said that there were a number of questions which had to be settled, and in his view they should only be discussed amongst the Bank, Treasury and Ministry of Supply. Certainly, at ministerial level the Governor, the Chancellor of the Exchequer, R.A. Butler, and the Minister of Supply would in due course come together for summit conferences to settle major points of principle as well as procedure. Their officials set up the Working Party on De-Nationalisation of Iron and Steel, which met for the first time on 5 November, the day before the King's Speech. At this meeting it was agreed that 'the first issue to be examined was that of the constitution, duties and powers of the body which would be set up to control the re-constituted free enterprise steel industry.... It was then necessary to consider just what assets were to be sold and in what form and parcels, to whom and at what price....'[15]

The Gilbert Committee, in fact, was setting out to cover some of the same ground as the bankers' Working Committee. Beale presumably would not have told them of this, since the Bank was not supposed to know about the Committee — and indeed, it was only on 6 November, the day after this meeting, that

14 The Minister of Supply had stated during the Debate on the Address on 12 Nov. 1951 that before introducing legislation the Government would consult the principal bodies representing the main interests affected, i.e. the Corporation, the BISF, the TUC and representatives of steel consumers. 493 H.C. Deb. 5s., col.574.

15 Cobbold to Bridges, 29 Oct. 1951, and P.S. Beale, 'Note for Record', 30 Oct. 1951, both File G1/125, Bank of England Archives. 'Note of A Meeting in Sir B. Gilbert's Room' [on 5 Nov.], 7 Nov. 1951, T.228/354/131995, Treasury Papers, PRO, London, for the quotation.

Peacock officially sent a copy of the Committee's Report to the Governor.[16] Forbes, for one, was alarmed by the plethora of committees being set up, and on 13 November he urged the other members of the bankers' Working Committee that

> the time had come when the Committee's recommendations should be more widely known. Duncan had told him that a small Committee had been set up of the Chancellor of the Exchequer, the Minister of Supply and the Governor, and they now knew that an official committee had also been set up consisting of very high Treasury officials who, as far as was known, did not know about this Committee. The Treasury committee had consulted two members of the Bank of England, who appear to have very little knowledge of the subject. Forbes also thought that it very likely that in addition the Minister of Supply would be setting up his own Committee.

In short, there was some danger that all of these committees would work independently, and would waste time considering problems which the bankers' Committee had already been considering for nine months.[17]

These fears were conveyed to the Governor by Peacock and Forbes, and when the Governor met Butler on 17 November he told him something of the City talks; the Chancellor 'thought it would be useful to explore further and would be grateful for anything we could suggest.'[18] Thus when Peacock and Bicester called to see the Governor at his request on 20 November, he suggested, and they agreed, that it would be useful to continue the work begun by the Working Committee, but now under the auspices of the Bank. It was agreed that this new Bank of England Working Party should be chaired by H.C.B. Mynors, an Executive Director of the Bank; the members would include Hugh Kindersley, managing director of the merchant bank Lazard Bros. and a non-executive Director of the Bank (who, with Mynors, would represent the Governor and Deputy Governor), as well as Erskine of Morgan Grenfell, Phillimore of Barings, Forbes of the FBI and Morison. Their task would be to advise the Governor and Deputy Governor on any recommendations they might make to the Chancellor, although the Governor assured Peacock and Bicester that he would consult them before he discussed

16 Peacock to Governor, 6 Nov. 1951, Partner's File 65A, f.95, Barings Archives.

17 Working Committee, 13 Nov. 1951, Partner's File 65A, f.98, Barings Archives.

18 Governor's Note,19 Nov.1951,File G1/125,Bank of England Archives.

the Working Party's recommendations with Ministers. On 23 November the Governor told the Chancellor, and then Bridges and Eady, of the existence of the Working Party; Mynors told Gilbert, who offered to see him whenever he liked and to give all the help he could. Both Chancellor and officials continued to welcome the help of the bankers, although the Minister of Supply was later to be annoyed at being denied knowledge of their individual identities.[19]

Meanwhile, the Cabinet were trying to establish their own chain of command, a task complicated by the fact that it was deemed impolitic to leave steel wholly to the Minister of Supply, Duncan Sandys. The Prime Minister set up the Steel Committee, to which Sandys had to report; one observer ascribed this arrangement to the fact that Sandys had many detractors, who believed he owed his advancement largely to the fact that he was Churchill's son-in-law.[20] However, the fact that Sandys was not a member of the Cabinet (nor had Strauss been) was probably part of the reason. Arrangements were not sorted out until the third week in November. Crookshank, who had chaired the original Ministerial Committee (which had decided that steel legislation was not possible before Christmas), wanted to shuffle off the responsibility. Attempts were made to get Lord Leathers, the Co-Ordinator of Transport, Fuel and Power, to take over, but he refused, so Crookshank remained the chairman, with Lord Swinton, the Chancellor of the Duchy of Lancaster, and Oliver Lyttelton, the Secretary of State for the Colonies, the most active members of the Committee. Mynors of the Bank thought that 'it seems likely that the main running at Ministerial level will be made by Duncan Sandys, using Swinton for advice and reporting to Crookshank as and when necessary.'[21] Certainly the man most closely concerned with the drafting of the Bill, the

19 Governor's Note, 23 Nov. 1951 and H.C.B.M., 'Steel', 23 Nov. 1951, both File G1/125, Bank of England Archives. Figgures passed on Sandys' request that he should be informed of the constitution and the names of the Bank of England Working Party, but as the latter were 'extremely anxious' to remain anonymous and the Governor supported them, the Treasury felt unable to pass the names on. Elkinton to Lindsell, 12 May 1952, BT 155/120, No.31, Board of Trade Papers, PRO, Kew.

20 Harold S. Kent, *In on the Act: Memoirs of a Lawmaker* (London: Macmillan, 1979), p.234.

21 H.C.B.M., 'Steel', 23 Nov. 1951, File G1/125, Bank of England Archives.

Parliamentary Draftsman, considered that Sandys had unlimited drive and capacity for hard work as well as considerable political skill. (He also thought him the most exacting and exasperating Minister for whom he had ever worked.)[22]

The Cabinet decided that steel denationalisation would constitute one of the three major bills for the coming session (the other two being bills to denationalise long-distance road haulage and to set up a Monopolies Commission), and it should therefore be ready for introduction soon after Parliament reassembled at the end of January 1952. Sandys was anxious to get it through in the current session, which meant having a Second Reading before Easter; after that the congestion caused by the budget and attendant financial business would render the introduction of a bill of any complexity or political importance virtually impossible. Accordingly, pressure was put on the officials to sort out the general principles and get a draft on paper.[23]

The Gilbert Committee (officials from the Treasury, the Ministry of Supply and the Bank) had met on 5 November, and by 7 November the Ministry of Supply had drafted a Note setting out the constitution and powers of a new Iron and Steel Board. An interim Report was ready for submission to Ministers within a very few days, but it was not discussed by Crookshank's Steel Committee until 22 November, since it was only then that the membership of the Committee was decided. Letters now went out to the BISF, the TUC, the Association of Iron Ore Producers and the Iron and Steel Corporation, inviting them all for discussions; the BISF had already prepared a paper, which was sent in to Gilbert.[24]

Designing a Steel Board was largely a task for the officials, since Whitehall contained within itself vast experience of government regulation of industry. Advice was sought from interested parties and acted upon, but on the whole the Ministry of Supply set about this task with confidence. However, determining how to return the industry to the private sector was another matter: Treasury officials hardly stood around looking

22 Kent, *In on the Act*, p.235.

23 Cab. 10(51), Minute 1, 22 Nov. 1951, Cab. 128/23. Kent, *In on the Act*, p.235.

24 'Iron and Steel De-Nationalisation: Constitution and Powers of a New Board', 7 Nov. 1951, T.228/354/131995. 'Working Party on De-Nationalisation of Iron and Steel', 9 Nov. 1951, and H.C.B.M., 'Steel', 23 Nov. 1951, both File G1/125, Bank of England Archives. BISF Council Minutes, 18 Dec. 1951, Box 01546, BISF Papers.

helpless, but they welcomed the aid of the Bank and the City. The bankers' Working Committee had drawn up a set of principles which they thought should govern the operation, as well as setting out, in general terms, some of the procedures. But a Bill required firm proposals and some detail, and this was what the Bank's Working Party was set up to provide.

It had been assumed from the outset by all concerned that an organisation separate from a statutory Board would be set up to deal with the return of the firms to private ownership, but its precise form had not been decided. On 19 December 1951 the Governor, after consultation with Bicester and Peacock, met with the Chancellor and the Minister of Supply to convey the Working Party's suggestions. First of all — and its prominence was revealing of bankers' fears — the Governor urged the adoption of a guarantee clause[25] in the Act as well as in the future prospectus. Secondly, the Board should control prices and development, but the Minister through the Board should not have the power to fix prices arbitrarily. Thirdly, a Realisation Company should be set up to hold and manage the nationalised companies as well as to sell them back to the private investor; the responsible department should be the Treasury (rather than the Ministry of Supply), which should also assume responsibility from the Corporation for the eventual repayment of the Steel Stock. The Governor emphasised that it was impossible to regard the operation as purely unscrambling: previous owners ought not to have any definite rights, although in most cases efforts should be made to return securities in privately-owned companies to former holders and some priority (e.g. by the use of pink forms) should be given to former holders of securities of publicly quoted companies (i.e. those listed on the Stock Exchange). The City still could not agree on the basis of the selling price: 1949 takeover price or market price at the time of sale? At this point there was a slight preference for the takeover prices plus something for accrued profits, but the Governor urged that this should be left as open as possible in the Bill. Finally, the Governor and his advisers thought the objective should be first to dispose of the private companies, and then in one operation dispose of the main block of the largest previously publicly-quoted companies.

The Governor did his best to impress upon the Ministers the importance of climate for the operation, both the political climate

25 The clause should somehow guarantee the investor against loss if the industry was renationalised.

— and here he urged political compromise with the Opposition, if possible — and the market climate at the time of sale. Further, neither the Government nor the City must have their hands tied by the Bill: details must be left open. 'It must be recognised privately that a big operation may not prove possible on any reasonable terms, and H.M.G. must not be committed to something they cannot do when the time comes.' At any rate, a great deal of work had to be done if the operation was to take place as soon as the legislation was passed, and this required full information from the existing Iron and Steel Corporation. He also thought that when the Government had a clearer idea of their intentions, some leading institutional investors ought privately to be sounded.[26]

To the Governor, it appeared that the Ministers agreed with the main lines of his (and the Working Party's) arguments.[27] As it happened, there was a large measure of common ground, but on two important points the Bank lost: one was the question of the type of organisation which would hold and market the companies, the other was the mooted guarantee clause. The City wanted a Realisation Company, while from the beginning the Treasury referred to a Realisation *Agency*. A company would be more independent than an agency, which would be 'merely the agent of, and therefore subordinate to, a Minister.' But the Treasury argued, and the Ministry of Supply agreed, that because the Realisation Agency would be a holding (and therefore to some extent an operating) organisation, since time would elapse before the companies could all be sold, and because until that time the assets were owned by the State, 'some degree of subordination to the Minister there must be. The Bank more or less accepted this', but then attacked the Treasury's flank: there were, the Bank

26 Working Party, 18 Dec.1951,Partner's File,50A, f.79, Barings Archives. C.F.C., 'Iron and Steel Act: Notes for Discussion with Chancellor of the Exchequer and Minister of Supply', 18 Dec. 1951, Board of Trade Papers, BT 255/120/131958, for the quotation. The Prudential Assurance Company had in fact already been sounded out, with somewhat depressing results: they were reluctant to contemplate underwriting a substantial amount. Working Party, 11 Dec. 1951, Partner's File 50A, ff.56–7, Barings Archives. Possibly the Governor did not know about this approach, since in his talk on 19 Dec. with Butler and Sandys, 'I mentioned that we had not as yet made any approach to institutional investors.' C.F.C., 'Iron and Steel Act', 19 Dec. 1951, BT 255/120/131958.

27 Working Party, 14 Jan. 1952, Partner's File 50A, f.88, Barings Archives.

argued, 'obvious possibilities of embarrassment' in any arrangement whereby the Agency (responsible for the 'operation' of a part of the industry) and the Board (reponsible for general policy supervision for the whole of the industry) were responsible to a single Department and had possibly conflicting loyalties to one Minister. Therefore, the Bank suggested brightly, while the Board would properly be responsible to the Minister of Supply, the Agency, whose duties would be essentially financial, should be under the aegis of the Treasury[28] —and as far from an interfering Department as possible. The Treasury, from the top to the bottom, objected to this. They would have to set up an Iron and Steel Department; the Agency's holding functions were likely to be longer-lasting than the Bank implied, and the Treasury knew nothing about the steel industry; and the Financial Secretary to the Treasury and the Chancellor objected to having to answer in the House of Commons for the steel industry when their department would have no real influence.[29]

But the Treasury lost this particular battle. A compromise was worked out (Clause 22 of the second draft of the bill) whereby the Treasury was to give directions to the Agency in all matters relating to the sale and the Minister of Supply was to give directions in all matters relating to holding. But then Sandys decided that he 'should have nothing to do with it'. The Governor had warned that the steel companies might not be saleable if the City thought that the Minister of Supply was able to interfere in the negotiations between buyer and seller, and this warning gravely disturbed Sandys; beyond that, he did not want day-to-day contact with the industry, but to be involved at a distance through

28 Minute to F. Figgures, 31 Dec. 1951, T. 228/354/131995.

29 *Ibid.* Minute by Figgures for Gilbert, 3 Jan. 1952, Figgures to Gilbert, Allen and Armstrong, 23 Jan. 1952, and J.A. Boyd-Carpenter to the Chancellor, 24 Jan. 1952, all T. 228/354/131995. 'The Chancellor of the Exchequer accepted the draft Bill with some reluctance because the main responsibility for the Agency was now placed on the Treasury.' Cab. 32(52), Minute 3, 20 March 1952, Cab. 128/24. The House of Lords would provide its own mite of embarrassment, Lord Wilmot pointing out during the Second Reading debate on 31 March 1953 that 'As to who will constitute this Agency, we are left in complete doubt. They are people who are going to be appointed by the Treasury. They are not even to be appointed by the Ministry of Supply, which has acquired great knowledge and experience of this industry. The Treasury is concerned purely with financial circumstances, yet, during all these years, it is this little body of Treasury appointees who are going to administer the steel industry....' 181 H.L.Deb. 5s, col.368.

the Iron and Steel Board.[30] The combined pressure of the Bank and the Minister of Supply defeated even the Treasury, and it became the responsible ministry for the Holding and Realisation Agency.

The other argument which the Bank and the City lost was over a guarantee clause. They had powerful allies — in particular Lord Swinton — but the sheer impracticality of the proposal was decisive. Ever since Strauss had pledged on 12 November 1951 that Labour would renationalise steel, Peacock and others in the City had worried that this statement rendered the whole proposed operation impossible. Things were made worse, Peacock believed, by the lack of an immediate response on the part of the Government.[31] Consequently, the insertion of a guarantee clause in the Bill, by which either the Treasury or the Agency would undertake to re-purchase the shares if renationalisation threatened, seemed to the City to be the only solution. Swinton was made aware of the City's fears, and on 10 December 1951, nine days before the Governor proposed such a clause to the Chancellor and Minister of Supply, Swinton wrote to the Solicitor-General, Reginald Manningham-Buller, for his professional advice on the matter. The Solicitor-General was entirely discouraging: 'For the giving of the undertaking to be effective it must be given in such a fashion that it cannot be avoided or repudiated by a succeeding Socialist Government. I have thought a good deal about this and I must confess I can think of no way in which such an undertaking can be given....'[32]

It might be thought that the considered advice of the Solicitor-General would have scotched the attempt to produce a guarantee clause — certainly the Treasury thought it had — but it did not. As Sir Bernard Gilbert minuted on 23 January 1952,

> At the meeting of the Steel Committee last night there was a great deal of discussion on the correspondence between Lord Swinton and the Solicitor-General on the giving of a guarantee that if when they came to power a Socialist Government nationalised iron and steel once again the shares would be bought back by the Government at some price or on some basis stated in the guarantee which would be endorsed on each share certificate. I thought that the

30 Figgures to Gilbert, Allen and Armstrong, 23 Jan. 1952, Treasury Papers, T. 228/354/131995.

31 Working Party, 11 Dec. 1951, Partner's File 65A, f.102, Barings Archives.

32 Manningham-Buller to Swinton, 8 January 1952, T. 228/354/131995.

Solicitor-General's arguments had effectively disposed of this proposal, but it was evident last night that the other Ministers (except the Minister of Supply) did not think so. They seemed convinced that without such a guarantee the shares would not be disposed of at all, and they answered the argument that a Socialist Government would repudiate the guarantee by saying that when it came to the point they would appreciate that the repudiation of an obligation entered into by Parliament would have such a serious effect on Government credit generally that they would not in fact repudiate.

We then went on to a very long discussion about the form which the guarantee should take, and in the end I said that I would see if I could put some considerations on paper.[33]

In early February the current draft of the bill contained a form of guarantee, and there would be other such attempts. As far as the Parliamentary Draftsman was concerned, the attempt wasted much precious time and was a major cause of the failure to get a bill ready by early February, as Sandys had wished. In the end, the obvious had to be faced: constitutionally, no Parliament can bind its successor and no form of words was going to bridge the gap between City wish and hard reality.[34]

Meanwhile, the drafting of the bill as a whole had gone on apace, and the first agreed version was submitted to the Cabinet by Crookshank on behalf of the Steel Committee on 28 December 1951. At this point the main provisions of the bill provided for (1) the winding up the Iron and Steel Corporation, (2) the transferring of ownership of the steel companies to a Realisation and Management Agency, and (3) the supervision of prices and development in the industry by a Board which would include outside members as well as representatives of the industry, trade unions and consumers. The Cabinet took note of the Report,[35] and the Departments, the Ministers and the Parliamentary Draftsman continued with their consultations and drafting.

The Bill went through at least seventeen drafts,[36] but the care taken in its preparation was an important reason why its eventual passage through Parliament would be easier than might have been expected. For example, it took a month, during which several memoranda were written and consultations held both with the Joint Iron Council and the BISF, just to decide upon a

33 Minute by Gilbert to Eady, 23 Jan. 1952, T. 228/354/131995.

34 Kent, *In on the Act*, p.236.

35 'Interim Report of the Steel Committee', C(51)56; Cab. 20(51), Minute 5, 28 Dec. 1951, Cab. 128/23.

36 At this point the author stopped counting.

definition of the industry: was it to be the IDAC definition, or all the sections subject to control during the War?[37] The Parliamentary Draftsman left in his memoirs a graphic description of what he and the officials went through in trying to produce the Bill:

> Early in January 1952 I got some draft clauses into print, and it was to discuss these that I attended my first conference with the Minister.... Socialist Ministers love bureaucrats, and Conservative Ministers regard them with deep suspicion, and none more so than Duncan Sandys....
>
> My draft clauses were so tentative that I did not think that they were worthy of the Minister's attention. I hoped to discuss a few crucial problems and then prepare a more comprehensive print. I thought that the conference would last two hours at the most, but at the end of that time we had just finished clause one. When we finally adjourned at 1.30 a.m. we were just about half way through the print, and were to resume at nine. The Minister's method was to go through the document line by line and indeed word by word. It was fatal to suggest that a particular subsection was simple and raised no problems; that was the signal for an even closer scrutiny. When Rowlands said that we had disposed of all points of substance and that the wording could be left to the draftsman, that was equally counter-productive....
>
> We went on as we had begun, with interminable conferences in the Minister's room.[38]

Meanwhile, the Bank's Working Party was considering the problem of the Realisation Agency in greater detail, trying to decide both what it would be and what it would do. At the same time the members attempted to ascertain sentiment in the City. After several meetings, Mynors drew up a report which was circulated only amongst the members of the Working Party and the Governor. It was then that the approach was made to the Prudential Assurance Company to sound them out on their reaction to the proposed resale. The response was quite discouraging, with the company indicating reluctance to participate to any great extent in the underwriting of an issue, as long as there was no agreement between the two political parties over the future of the industry. The hope was also expressed that tender of Steel Stock at par, or close to it, would be allowed — a suggestion no doubt stimulated by the consideration that the prices of both equities and gilt-edged stocks had been falling since the election,

37 BT 255/108, No.2.
38 Kent, *In on the Act*, pp.234–36.

so that Steel Stock was trading at below par value. The inference the Working Party drew from this conversation was that no time should be wasted in drawing up an outline scheme and discussing it with representatives of the principal institutional investors, who clearly had to be educated. The other conclusion drawn was that the first public offering might have to be smaller than the Working Party had hoped.[39]

The report had formed the basis for the Governor's meeting with the Chancellor and the Minister of Supply on 19 December, but it was by no means in a final form. A version was ready by 31 December, but the members of the Working Party were still not entirely happy with it: the point of drawing up a report at all was to give the Governor something to send to Whitehall, and the Working Party feared that the Treasury or the Ministry of Supply might take as final suggestions which were only tentative. Another reason for their reluctance was that the revised memorandum 'made the possibilities look in most respects even more difficult than they were admitted to be and would not alter the present opinion (reported briefly by the Chairman [Mynors]) of Departmental officials that resale of Steel shares was not likely to be a practical proposition.'[40] Figgures wrote to Gilbert in an internal Treasury minute that 'in the Ministry of Supply more and more thought is being given to alternative forms of denationalisation in face of the increasing probability that straightforward denationalisation by sale will not be possible.'[41] No reason was given — but perhaps they too read the share prices page.

Meanwhile, the Working Party had been told of the contents of the Steel Committee's Interim Report (the draft Bill), which had been submitted to the Cabinet on 28 December 1951. They were unhappy with some of the financial clauses, which envisaged that the Holding Company should be much closer to the Minister of Supply than they thought desirable, and these clauses were therefore reconsidered. But this may have stimulated the bankers to produce a final report for Whitehall, in order to make certain that their requirements were included in the final version

39 H.C.B. Mynors 'Report', 10/11 Dec. 1951, ff.36–37 and Working Party, 11 Dec. 1951, ff.56–57, both Partner's File 50A, Barings Archives. Francis Whitmore, 'Problems of Steel Share Unscrambling', 28 Jan. 1952, *Daily Telegraph*.

40 Working Party, 14 Jan. 1952, f.88, for the quotation, and H. Mynors, 'Memorandum', 25 Jan. 1952, f.89, both Partner's File 50A, Barings Archives.

41 F.E.F., 'Iron and Steel', 2 Jan. 1952, T.228/354/131995.

of the Bill. By 25 January 1952 the revised report was ready, and the Governor sent copies to the Chancellor and the Minister of Supply.[42]

Whether and when these two Ministers actually saw the report is open to question, however. The Treasury officials apparently agreed that the report dealt with questions which were properly matters for the Agency,

> and while it is most desirable that the Bank and the Treasury should discuss them in advance it seems to me most unwise to give them a wider circulation. On the evidence of what has happened to date, it seems not unlikely that the substance of the documents would be in one of the financial columns shortly after its circulation to the Steel Committee.[43]

By 29 April, two months later, the report had still 'never been put to the Ministers'; this was in spite of the fact that on 22 January the Cabinet Steel Committee had written to Sandys inviting him to report on his further discussions with the Bank of England about the resale of the industry. Sandys' Permanent Secretary then wrote to Gilbert to confirm an inter-departmental agreement that the Treasury should collaborate with the Bank in preparing a report giving the substance of the Working Party report.[44] It can only be assumed that the Treasury and the Bank agreed to produce this with all deliberate speed, since the Bank would share the Treasury's horror at unguided comment in the newspapers. However, it is probable that some version of the report was seen by Sandys in early May, possibly in preparation for the Cabinet meeting on steel on 7 May 1952, because he wanted to know the names of the members of the Working Party — and the Bank and the Treasury refused to tell him.[45]

The report, then, appears to have been kept within the two Departments, where officials used it in drafting the Bill, which would then in any case be gone over by the Ministers (although there is not much evidence that, in the face of his other duties, the Chancellor was particularly interested in the Steel Bill). The report again emphasised the size and complexity of the operation, the fact that the political threat was outstanding, and the point that the operation was much more than the reinstate-

42 'Iron and Steel', 25 Jan.1952, Partner's File 50A,f.94, Barings Archives.
43 F.E. Figgures, Minute to Gilbert, 1 March 1952, T.228/355/131995.
44 Elkington to Lindsell, 29 April 1952, BT 255/120/131958.
45 Elkington to Lindsell, 12 May 1952, BT 255/120, No.31.

ment of former owners. It would be impossible to indicate either the timing or the size of the operation until the legislation was through Parliament, since market conditions were impossible to predict. Arrangements should be made for a substantial part of the industry to remain in State hands for an indefinite period. The objective of the whole exercise was the well-being of the industry as a national asset and particularly its power to raise capital for future investment — not the financial results to the Treasury nor the interests of investors. With regard to the powers of the proposed Iron and Steel Board, which would be concerned especially with prices and development, investors were probably ready to accept control on the scale proposed, since it had operated for some time past, although less formally. However, the Board should have no responsibility for the executive of the industry and should act as an independent and responsible body on prices, with the retained power of the Minister a reserve power. The Holding and Realisation Agency would be primarily a holding and marketing body, and Ministerial control of the Agency should be as remote as possible. Finally, with regard to the eventual price to be set for the sale of the shares, the Working Party considered that the most satisfactory starting price would be the 1949 take-over value, with an additional amount added (a 'loading') to cover the expenses of the operation plus the profits accumulated since the vesting date.[46]

The Departments took the report quite seriously, and made a number of changes in the bill to accommodate the points of the Working Party (and, in a number of similar cases, those of the BISF). For example, in response to the point that the amount the Treasury received for the shares should not be as important a consideration as the financial health of the industry, the Treasury removed the words 'in the interests of public finances' from paragraph 24. (The BISF independently pushed for this change.) In response to paragraph nine of the report, which stated that it must not appear that both the Agency and the Board were to have control over the industry, and that Ministerial control must be remote, Section 25 of the bill was re-drafted; in addition, it was planned that a Government spokesman would make the point at the time of the Second Reading that the Agency would not be expected to interfere in day-to-day business.[47]

46 'Iron and Steel',25 Jan.1952, Partner's File 50A,f.94, Barings Archives.
47 'Summary and Comments on the Bank of England Report', 11 Feb. 1952, BT 255/120/131958.

These attempts to accommodate the wishes of the City were on the whole successful. On 11 February Mynors gave a copy of the draft Iron and Steel Bill to Phillimore of Barings and to the other members of the Working Party, and on 14 February they met to consider it. The Working Party accepted that the potential investor would probably be content in general with the powers given to the Board and to the Minister with regard to price control, although not perhaps in one or two other respects.[48] Attempts would continue to be made to modify these clauses in the Bill, but for the time being, the formal activity of the Working Party was finished.

The Cabinet Crisis Over the Bill

On the same day as the Working Party met to consider the draft Bill, the Cabinet met to consider what to do with it, as well as other legislative plans, all of which had suddenly been overturned. The death of King George VI had been announced on 6 February, necessitating an adjournment of Parliament and consequent loss of time. Unfortunately, no decision could be taken at this Cabinet, since the Bill was not yet in final form; once it was, the Cabinet would decide how best to handle it in Parliament.[49] To beat the Bill into final form, however, took another two months. The Parliamentary Draftsman believed that

> we might have had the Bill reasonably fit for introduction by the end of February but for the Steel Committee. They had to be consulted at every stage, and the draft eventually prepared for submission to the Cabinet had to be approved by them. It was not possible to bulldoze the Bill through these seasoned politicians, although the Minister certainly did his manful best to do so.[50]

The Steel Committee finally agreed on a draft Bill and it was discussed in Cabinet on 20 March 1952. The main shape of the Board remained broadly as it had been when described to the Cabinet in late December, but more details were given about the Agency, which would have wide powers to dispose of the industry either by sale or leasing. Disposal by leasing was a new addition. Sandys argued that it might be necessary to make substantial use of leasing, because this would reduce the capital which

48 Working Party,14 Feb.1952,Partner's File 50A,f.97,Barings Archives.
49 Cab. 17(52), Minute 3, 14 Feb. 1952, Cab. 128/24.
50 Kent, *In on the Act*, p.237.

private enterprise would have to find — clearly a reaction both to the reluctance of the institutions and to falling market prices. The reaction of the Cabinet was, on the whole, favourable; the one notable exception was the Chancellor. He pointed out that since it had not been found practicable to include a provision guaranteeing investors against further nationalisation, it would probably take a long time to dispose of all the companies. The Treasury might therefore find themselves *de facto* owners of a large part of the iron and steel industry for a considerable time, while being unable to dissociate themselves from the companies' dividend policies or from the need to provide finance for development. 'He realised that the Agency would be expected to devote its attention primarily to the disposal of the industry and to leave the individual companies to run their affairs with the minimum of interference; but, despite this, the responsibilities assigned to the Treasury were potentially embarrassing.' But however reluctantly, he gave his approval to the draft Bill, and the Cabinet approved it in principle, inviting the Steel Committee to take account of any further comments or objections, and to report to the Cabinet in a fortnight's time.[51]

The legislative career of the Iron and Steel Bill was now interrupted by two obstacles: time, and the Prime Minister. By the time the Bill went back to the Cabinet in mid-April, after the Budget, it was probably already too late for it to get through that Session. Secondly, Churchill decided to take a personal interest in both iron and steel and transport denationalisation, and he decided that transport should take precedence. It is not clear just how devoted Churchill was to steel denationalisation in its own right. According to his biographer, 'the Labour Government's intention to nationalize the steel industry seemed to Churchill [in the autumn of 1947] to offer a focal point for combined Conservative and Liberal action',[52] and this was a theme that remained important through the next election: in March 1950 he proposed that a committee of the Shadow Cabinet be set up to 'go into all the questions open between Conservatives and Liberals and to see what can be done to secure greater unity among forces opposed to Socialism'; the only specific topic he listed was opposition to steel nationalisation.[53] In short, his original oppo-

51 Cab. 32(52), Minute 3, 20 March 1952, Cab. 128/24.
52 Martin Gilbert, *Winston S. Churchill. Volume VIII: 'Never Despair', 1945-1965* (London: Heinemann, 1988), p.346.
53 *Ibid*., p.529.

sition to nationalisation seems to have been tactical rather than strategic. This is not to argue that he did not believe steel should be denationalised: certainly in January 1952, the Parliamentary Draftsman recalled, 'such Churchillian phrases as "striking off the Socialist shackles" and "lifting the burden of bureaucracy from the backs of the People" were floating about.'[54] But he also believed that long-distance road haulage should be denationalised — at lunch on 19 April 1952 he proclaimed that 'This is a great issue, more important even than steel'[55] — and with time short, a choice would have to be made as to which would take priority for that Session.

A number of Conservative backbenchers, however, failed to see why a full-blooded denationalisation bill for iron and steel as well as for road haulage could not go ahead immediately, and in the first week of April 1952 a motion critical of the Government's slow progress was put down by Cuthbert Alport, which collected forty-six signatures.[56] On 9 April, in answer to a Parliamentary Question, the Prime Minister said that the Government intended to denationalise road haulage at the earliest possible moment; he also referred to their intention to denationalise iron and steel, but the impression was certainly left that road haulage was viewed by the Government as the more urgent of the two measures. He apparently gave the same impression when he spoke to a meeting of the Conservative backbench 1922 Committee later that day, although here he declared categorically that the Government intended to pass both Bills during the current Session, even if it meant prolonging the Session until after Christmas. As soon as Hunter, the President of the BISF, had heard reports that there was pressure for the Steel Bill to be held back in favour of transport, he wrote to Sandys to emphasise the damage which would be caused to the industry by further delay.[57] (Andrew Duncan had died suddenly earlier in the month, and Hunter now conducted most of the negotiations on behalf of the

54 Kent, *In on the Act*, p.233.

55 Lord Moran, *Winston Churchill: The Struggle for Survival 1940–1965* (London: Constable, 1966), p.385.

56 Anthony Seldon, *Churchill's Indian Summer: The Conservative Government 1951–55* (London: Hodder and Stoughton, 1981), p.187. Herbert Morrison said to Richard Crossman that 'the Tory backbenchers were playing the same disruptive role in Parliament as the Bevanites.' Morgan, ed., *Backbench Diary*, entry for 10 April 1952, p.98.

57 Notes for meeting of the Executive Committee, 22 April 1952, Box 011612, BISF Papers.

BISF with members of the Government.) But Sandys could do nothing, and at the meeting of the Cabinet on 24 April, it was decided to drop the Steel Bill for the Session.[58]

The decision was made public by Churchill in a broadcast on 3 May, during which he indicated that transport legislation was to be given precedence over steel. The BISF were upset and alarmed, and responded immediately to Sandys' request for a memorandum for the Cabinet, setting out the reasons why the industry thought it was important that the Steel Bill not be laid aside. On 7 May 1952 the Cabinet met and decided to go ahead with the Transport Bill that Session, but not with steel. Sandys argued that its postponement would have serious consequences for the industry: the Corporation enjoyed neither standing nor authority, since it was soon to be swept away; further, the BISF did not represent the current ownership of the industry, and was aware that many of its functions would ultimately be taken over by the new Board. That meant that responsbility for decisions which ought to be taken by the industry would be thrown back on to the Government — not, of course, what the Conservatives had intended. Besides, he added, denationalisation ought to be concluded as long as possible before the next election.

No one apparently gainsaid Sandys' arguments about the industry, but he was challenged on the election issue: whatever the timetable, part at least of the industry would remain in public hands at the next election. The real problem was time. Unless the Bill could be examined in Standing Committee, rather than on the floor of the House, there was no possibility of its being enacted in the current Session. Yet, with a Government majority of only seventeen, the risk of sending such a highly controversial measure to a Standing Committee, unless special arrangements could be made for the replacement of Committee members who might fall ill, was thought to be too great to be faced. In the end, the Cabinet decided to postpone a decision on a timetable for the Bill. [59]

In the meantime, Hunter called on the Prime Minister on 12 May to impress on him the strength of feeling in the industry. He also played a political card, referring to 'the importance of the Government getting the benefit of the credit which was likely

58 Diary of H.F.C. Crookshank, 24 April 1952, f.66, Ms d. 361, Crookshank Papers, Bodleian Library, Oxford.

59 Chairman's Agenda for Executive Committee, 20 May 1952, Box 011612, BISF Papers. Cab. 50(52), Minute 6, 7 May 1952, Cab. 128/24.

to accrue later in the year from increased steel production — the Industry was now coming to the end of a difficult period of raw material shortages — and from the Second Development Plan.' The Prime Minister raised the difficult point of the parliamentary timetable, and this gave Hunter the chance to leave with Churchill a memorandum comparing the positions relating to steel and transport legislation: while it was known that there was still a lot of work to be done on the Transport Bill, the Steel Bill was in its final stages and virtually ready for publication. The Prime Minister naturally assured Hunter that he would consider the points he had raised.[60]

When the Cabinet met on 15 May, Sandys pointed out that because they had dropped the Steel Bill from the current Session, it would probably not become law before July 1953 and the Board would not be established before early 1954. In the circumstances he had two proposals for the industry: he would try to set up machinery similar to that provided for by the Bill, but on a voluntary basis, and he would set in train preliminary negotiations for the resale, as soon as the Bill became law, of those parts of the industry for which large investors were known to be holding capital available. (This probably referred to the companies which had been held as subsidiaries by the larger companies.) He then argued that all of these proposals, to which apparently no one objected, should be carried out against the background of public knowledge of the Government's intentions; he therefore thought it necessary either to publish the Bill and give it a Second Reading, although it could go no further in the current Session, or to embody the proposals in a White Paper, which could be debated in Parliament. Sandys himself preferred the second alternative. The Cabinet agreed with Sandys' arguments, but objected to publishing a Bill so far ahead of the time when it could become law. He was therefore asked to draw up a White Paper which would mean that the measure could be modified in the light of comments which came up in the debate. The White Paper, the Cabinet decided, could then be debated before the summer recess.[61]

Work began immediately on the White Paper, and meanwhile Sandys met with Hunter to discuss the proposals made in Cabi-

60 Chairman's Agenda, Executive Committee, 20 May 1952 for the quotation, and Notes for Meeting of the Executive Committee, 22 April 1952, both Box 011612, BISF Papers.

61 Cab. 53(52), Minute 5, 15 May 1952, Cab. 128/25.

net. He found that the BISF disagreed with most of them. Hunter thought unrealistic and undesirable the suggestion that an Advisory Committee might be set up as a forerunner of the Iron and Steel Board. He thought it unlikely that the unions would cooperate, and this might later prejudice the success of the Board. From the industry's point of view, he did not want 'yet another body on the Industry at a time when both the Ministry of Supply and the Corporation must participate actively in policy decisions.' As for the choice between a Second Reading or a White Paper, the BISF preferred the first alternative: it would leave the way open to complete the Bill if the Government were unable to produce the Transport Bill in time, and would in any case seem more solid evidence of the Government's firm intent than a White Paper. But what the industry would really dislike was the lack of any publication at all.[62]

Sandys' second suggestion in Cabinet had been that preliminary work should start on preparing companies for resale. This was the view of Morison and others in the City as well, and indeed Forbes told the Chancellor that if the operation was delayed for a year it would be 'quite impracticable'. The Treasury, however, accepted the Governor's view that it was unlikely to be more difficult in the future than it already was. Mynors predicted that once the White Paper was published, the Treasury would come to the Bank to ask whether it might be possible to set up some kind of 'shadow Agency' so that the decisions necessary in preparing for resale could be taken, if only unofficially. In view of these varying pressures, the Governor decided that as soon as the White Paper was published, he would re-convene his Working Party for advice.[63]

Meanwhile, the White Paper was drafted, and on 3 July 1952 it was presented to the Cabinet. Views on certain aspects had once again changed: for example, it was no longer proposed that the price of shares in the denationalised companies be based on the takeover price plus a loading; prices should instead be based on the current value of the assets they represented in the light of 'monetary conditions and the general level of security prices prevailing at the time' — in other words, based on market price.

62 Chairman's Agenda, Executive Committee, 20 May 1952, Box 011612 for the quotation, and Minutes of the Executive Committee, 20 May 1952, Box 01550, both BISF Papers.

63 H.C.B.M., 'Steel', 27 June 1952, with comments thereon, File G1/125, Bank of England Archives.

This decision reflected the Treasury's worries that by selling at the possibly lesser takeover value, the Government would be open to the charge of having sold off national assets at less than their true value (and to their friends at that). The Cabinet argued back and forth, and in the end referred the draft to the Steel Committee, while inviting Sandys to submit a final draft on 8 July. Crookshank then added that while he recognised that publication of the White Paper would reassure the Government's supporters and the industry, nevertheless he doubted whether their plans should be disclosed at once, when there was no chance of any legislation until the following year.[64]

This was the signal for a Cabinet battle in which the resignation of Lord Salisbury, the Secretary of State for Commonwealth Relations, was seriously threatened. Indeed, it can be argued that questions of timing exercised as much if not more emotion than questions of substance. Things started out promisingly. In Cabinet on 8 July the Prime Minister proposed that the White Paper be published, but that no debate be held on it before the summer recess. This would mean that the recess could begin early, and with the new timetable the Steel Bill could pass by the end of April 1953, two months earlier than had been anticipated. Sandys thought this was a very good idea: if it were debated before the recess, the Opposition would probably take up a position of 'uncompromising hostility', thus undermining all that Sandys had been working towards, which was as large a measure of cross-party agreement as possible. Indeed, he suggested that publication be postponed until just before Parliament adjourned, which would prevent the Opposition's using one of their Supply Days for a debate. The Cabinet then directed Sandys to present the White Paper — but on a date to be determined in consultation with Crookshank.[65]

64 Cab. 58(52), Minute 1, 10 June 1952, and Cab. 65(52), Minute 3, 3 July 1952 for the quotation, both Cab. 128/25. '...it is surely excessively dangerous to indicate in a document with a wide circulation that some regard must he had to the interests of the previous owners? While it may be necessary to do this in fact, it seems extremely undesirable to say so. Nothing is more likely to arouse political passion than the indication that some national assets may be sold for less than they are worth.' F.E. Figgures, Minute to Gilbert, 1 March 1952, T.228/355/131995.

65 The Transport Bill had also been put off, and under the Prime Minister's new plan would become law in March 1953. Cab. 66(52), Minutes 1 and 2, 8 July 1952, Cab. 128/25.

Unauthorised press leaks about the Government's intentions had accompanied the drafting of the White Paper, and battle was joined at the meeting of the Cabinet on 14 July 1952. Ministers' arguments were not minuted by name, but certain Ministers clearly took certain positions. Lyttelton claimed (in an interview some years after the event) that he had wanted denationalisation as early as possible. Crookshank and Woolton, the Lord President of the Council, wanted early denationalisation as well, and for it to be as thorough as possible; and they were presumably not very keen on all of the proposals for control by the State. On the other hand, Butler, Harold Macmillan, the Minister for Housing and Local Government, and Salisbury, approved of Sandys' concept, that the State retain some measure of responsibility and control, and with a real attempt being made to find a solution with which the Opposition, as well as the industry and the unions, could live. Lord Cherwell, the Paymaster-General, had reservations about denationalisation, and so did Leathers, although his was less a principled opposition than a tactical one — he feared a lack of purchasers. As for the Prime Minister himself, he blew hot and cold: while he had strongly supported denationalisation during the election, by September 1952 he was remarking to his crony Lord Beaverbrook that the sale of the companies was impossible and that the best solution was to leave control to private enterprise but ownership to the Government.[66]

At the Cabinet on 14 July, which was wholly devoted to steel, the Prime Minister announced that some Ministers were doubtful about the decision to publish the White Paper, when the remainder of its legislative career would have to wait until 1953. Cabinet opinion proceeded to divide sharply. Almost all of the arguments against proceeding at once with the Bill had a theme in common: fear of the Opposition and especially of union hostility. There were warnings of a serious sterling crisis which would require national unity to meet it; the Government had announced that the first priority was to redress an adverse balance of payments, with the goal a 20% increase in exports, and this would be threatened by a lack of union co-operation; and the Labour threat to renationalise would undermine the industry by threatening drastic changes with successive governments — it would be far better for the Government to find some basis of agreement within the industry and between the Parties in order to take it out of politics. Finally, what if, after all the political

66 Seldon, *Churchill's Indian Summer*, pp.190–91.

controversy, few buyers were found for the industry? It would be a serious blow to the Government's credibility.

Those who wanted to proceed with the Bill combined political points with economic arguments. Because the Government proposed to carry on with the Transport Bill to which Labour also objected, a call for national unity would be met with derision or ignored. True, to introduce the Bill would upset Labour, but not to introduce it would upset the Government's own supporters in and out of Parliament: the F.B.I., for example, had specifically asked that it be introduced without delay. The Party was pledged to it, and to abandon it would shake its cohesiveness. On a higher philosophical plane, to abandon the Bill would shake the confidence of all who believed in the principle of free enterprise, and indeed, there was a serious risk that all confidence would be lost in that principle unless the Government took 'active and early steps to foster it'. Finally, the argument over the best interests of the steel industry itself could cut both ways: the continuing uncertainty could harm its efficiency, since urgently-needed plans for development were being held up.

The Cabinet tentatively concluded that the best solution would be to find a basis for agreement — within the industry, if not between the Parties — and this would involve placing more emphasis on the Government's proposals for the future supervision of the industry as a whole and less on the proposals for denationalisation. Sandys said that 'he was not without hope that some basis of agreement could be found', and the Cabinet agreed that it would be a real advantage if the proposals commanded wide public support. But the Cabinet also agreed that the Government could not now wholly abandon its policy: the only question was timing and form. The Ministers then decided to continue the discussion the following day.[67]

The Prime Minister and Salisbury met with Hunter, who now changed his mind: if the Bill was to be introduced at the beginning of the new Session in the autumn, it was no longer essential that the White Paper be published before the recess, and he would be content with a firm declaration of the Government's intention to proceed with the Bill. In the Cabinet on 15 July, Salisbury revealed that Hunter's concession had relieved his mind: the doubts he had expressed in the previous day's Cabinet had related solely to timing; he did not want the Government to promise it for the near future, but he did not mind if they

67 Cab. 68(52), 14 July 1952, Cab. 128/25.

announced it. In the context, of course, this was a ridiculous position, and 'other Ministers pointed out that Hunter's view would not be met' unless the Bill was promised before the end of the year. The Prime Minister then suggested that if that was the case, the publication of the White Paper might after all be less provocative than a statement of the Government's intention in terms strong enough to meet the views of Hunter and the BISF. Added support for this view came with the admission that the leaders of the Parliamentary Labour Party had already learned 'through the usual channels' that the Government intended to publish a White Paper before the recess. The Prime Minister concluded that he favoured publication of the White Paper but without a debate on it before the recess.

The Cabinet still could not decide what to do, and temporised with the interim conclusion that they could not finally decide until they could see a draft of the statement to be made while introducing it. Harold Macmillan, David Maxwell Fyfe, the Home Secretary, and Sandys were sent away to draft the statement, while Sandys was instructed to revise the White Paper in order to lay even greater emphasis on the proposals for the future supervision of the industry and less on the proposals to return it to private ownership.[68] All of this was apparently in aid of pacifying Salisbury, whose loss to the Government because of his prestige, and influence on traditional Tories, would have been a great blow. Crookshank wrote in his Diary that in Cabinet that day there had been 'much argument about Steel. I still hope we can get a compromise & prevent Bobbety's [Salisbury's] resignation.'[69] These attempts continued the following day, as Crookshank again noted: 'Nothing much in [afternoon] except further conferences with Bobbety etc on Iron & Steel.'[70]

In the end, they were successful. When the Cabinet met again

68 Cab. 69(52), Minute 1, 15 July 1952, Cab. 128/25.

69 Crookshank Diary, 15 July 1952, f.76, ms d. 361, Crookshank Papers. Salisbury wrote to Swinton that 'I opposed Steel last night; but the broad view of those at the meeting was in favour of going on with it. They were, I thought, more acquiescent than enthusiastic. I remain opposed, & rather depressed. I find myself more & more out of sympathy with Winston, who seems to me purely out to reverse what has been done, instead of trying to create a new & healthier atmosphere of co-operation in meeting our difficulties.' This was handwritten on 10 Downing Street notepaper, presumably during Cabinet, but undated. File 174/6/1, Swinton Papers, Churchill College, Cambridge.

70 *Ibid*., 16 July 1952.

on 22 July, Salisbury announced that after looking at the statement drafted by Macmillan, Sandys and Maxwell Fyfe, he had decided that after all his ideas would be better met by the publication of the White Paper. He was particularly taken by Sandys' suggestions that after the issue of the White Paper, a statement could be made emphasising the Government's readiness to consider any constructive amendments to their scheme which could provide a basis for an agreed settlement. After this capitulation, there was general support in the Cabinet for early publication, and the Cabinet authorised Crookshank to announce, in his statement on Business in the House of Commons on 24 July, that the White Paper would be presented to Parliament the following week.[71]

On 28 July 1952, the White Paper on the Iron and Steel Industry was presented by the Minister of Supply to Parliament.[72] Press comment was generally favourable, and the Chancellor, in his speech on 29 July, referred to the Government's readiness to consider constructive criticisms (the terms of this reference were agreed with Salisbury first).[73] It was now time for the Parliamentary recess, and political controversy over the Bill temporarily died down, before warfare resumed with the debate on the White Paper in October. Meanwhile, after their own August holidays, the City began serious preparations towards the day when they would attempt to carry out the policy of the politicians.

71 Cab. 71(52), Minute 2, 22 July 1952, Cab. 128/25.
72 P.P. 1951–52, Cmd. 8619.
73 Cab. 74(52), Minute 1, 29 July 1952, Cab. 128/25.

3 PASSING THE BILL

From the publication of the White Paper in July 1952 to the announcement of the Royal Assent to the Iron and Steel Act in May 1953, the politicians and the civil servants concentrated on drafting, debating and legislating. At the same time the City, with the encouragement of the politicians, began the preparations necessary to bring the industry to market. Although this chapter will consider separately the activities of the politicians and the City, they reacted to and against each other. In the first section Ministers and the City consider how to proceed in the period before the Bill became law: they were all agreed that time should not be wasted. In the second section the concentration is on political developments, as civil servants continually re-drafted the Bill and the politicans attempted to get it through Parliament. In the third section the focus is on developments in the City, as work secretly began on the preparations, so that once the Bill was passed the resale could take place without delay.

Summer in the City

The period between early August 1952 and the debate on the White Paper on 23 October saw more questions raised than answered, as the politicians, the Issuing Houses and the industry's leaders all tried to plan the way forward. On 1 August 1952 Mynors of the Bank, Phillimore of Barings, Forbes and Morison met to consider how things stood now that the White Paper had been published. The general opinion of the group was that the events of the previous few months had probably made the operation more difficult, but not impossible. What clearly continued to haunt the Working Party was the fear that the institutions would be unwilling to allocate enough funds to the purchase of steel industry securities. Morison thought it essential to find ways to reduce the amount of money needed from the market. The Treasury, he argued, should be prepared in the case of each company to hold up to 75% of the resale value in the form of prior charges,[1] bearing a relatively low rate of interest, an idea

1 Prior charges are securities, either quoted or unquoted, where the

which, in modified form, would later be adopted. The Treasury, in other words, might need to be called in to redress an unwillingness of the institutions to take up sufficiently large allotments of ordinary shares; and certainly, a threnody which wove its way through months of discussion was a lament for the institutions' collective timidity. Might the solution, the group wondered, be to extend the Working Party to include representatives of insurance companies and investment trusts so that large investors might be brought into the picture as early as possible?[2] Apparently not, and the institutions remained terrified of the political risks.

Finally, the Working Party discussed who should sponsor the offers for sale. The merchant bankers continued to hope that it would be the Bank of England itself, since the Bank was the channel by which Government financial business had traditionally been done, but Mynors attempted to crush that idea. The Bank, he stated, would not consider making the offers for sale, although they would of course help — and indeed, their participation would be essential, since one way of paying for the shares would be to convert or switch Government Stocks (i.e. gilt-edged stocks) for them. But he (and presumably the Governor) imagined that there would be syndicates of Issuing Houses for each issue, the leadership of each depending on which House had previously handled the affairs of the company concerned.[3] In fact, that was precisely what happened, and partly accounted for the leading position of Morgan Grenfell once the technical preparations for the resale commenced.

A week later Mynors wrote to the Treasury, suggesting that it might be time to begin the preparation for offers for sale of one or more of the companies. These preparations were expected to be very time-consuming, since they entailed going over the accounts and reports of recent financial years, as well as trying to determine what the profits would have been if the companies had remained in private hands. The Treasury agreed that 'this would be valuable', and the Ministry of Supply were willing to provide whatever information they had to help in the exercise. The Treasury, however, refused to commit the Government to

capital and interest have priority over the ordinary (equity) shares. They include preference shares, debentures and loan stocks.

2 Working Party, 1 Aug. 1952, Partner's File 50A, f.105, Barings Archives.

3 *Ibid.*

any definite views on whether the sale price should be takeover price or market price, telling the Bank to assume that they should aim at the 'best available proceeds of sale'.[4]

The Minister of Supply, too, was thinking of the future, although his horizon was somewhat more circumscribed. On 14 August he wrote to the Governor:

> Now that the White Paper on the Iron and Steel Industry has been published and, on the whole, been well received, we should, I think be clearing our minds on the arrangements to be made for disposing of the nationalised companies.
>
> In the debates which will take place when Parliament reassembles in October we will almost certainly be pressed for information about our intentions. We will, of course, say as little as possible but we should have some plans for the operation — at least in embryo.

He ended by suggesting that he and the Governor should meet upon Cobbold's return from his holiday.[5] They did so a week or so later, and Cobbold told Sandys that he would do 'a little exploration in the City' and talk to him the last week in September.[6]

Sandys also contacted the leaders of the BISF, meeting Hunter in mid-September to tell him how satisfied he, Sandys, had been with the reception of the White Paper. 'This had been much more favourable [according to Sandys] than that on the Government's Transport policy — in particular, there had been no immediate hostile reaction from the Parliamentary Labour Party or from the Industry's Union.' He then assured Hunter that the Government welcomed any constructive criticism of their proposals. The Council of the BISF considered this, but had no further suggestions to offer, 'apart from emphasising the importance of the Government taking whatever action might be possible to counteract any re-nationalisation threat.'[7] Sandys, of course, had been attempting from the beginning to construct a Bill which would minimise just that.

All now seemed to be agreed that it would be a good thing for work to begin on the technical problems of denationalisation. The question was how. Mynors of the Bank had predicted

4 Treasury to Mynors, 15 Aug. 1952, T.228/356/131995.
5 Sandys to Cobbold, 14 Aug. 1952, File G1/125, Bank of England Archives.
6 Cobbold's Note, 27 Aug. 1952, on *ibid.*
7 BISF Council Minutes, 16 Sept. 1952, Box 01546, BISF Papers.

in June that once the White Paper was published, the Treasury would come to the Bank to ask whether it might be possible to set up a shadow Agency, and now they did. On 5 September 1952, in a conversation with D.G.M. Bernard, the Deputy Governor of the Bank, E.G. Compton, Third Secretary to the Treasury,

> mentioned that the Steel Bill would be coming forward in the new Session and he had been asked to raise with us the question of the proposed Steel Agency for holding the shares. The idea had been to form a Shadow Agency before the Bill became law so as to study the problems and be ready to proceed to business when the Bill became law. The Treasury had been waiting to hear from the Ministry of Supply what they thought of this: they had now heard that they were in favour of it. Hence Compton asked us if we were still in favour of the shadow Agency and if we would suggest names.[8]

If the Bank had once been in favour of such a shadow Agency, it was no longer — or at least not in favour of a public one. The Governor wrote on these lines to the Chancellor on 11 September, and the two of them, plus the Minister of Supply, met on 16 September to sort out what should be done. The Bank was opposed to an Agency or to 'shadow appointments' to the Agency 'until the Parliamentary position made it appropriate', but it was in favour of the necessary work being put in hand. The approach lighted on was to invite a leading accountant 'such as Sir John Morison' to assist the Treasury and the Ministry of Supply, 'without anything being said between H.M.G. and Sir John Morison about the eventual Chairmanship.' The three agreed that he should have an inter-departmental committee and secretariat to help him, and that he would chair the committee. In short, there would not be a shadow Agency, but there would be an 'unofficial' shadow Chairman and support staff. The Governor had already sounded out Morison on the subject, and he would now have another word with him in general terms.[9]

They met two days later, and the Governor told Morison that he was proposing to the Government that they should forthwith put him in charge of the shadow Holding and Realisation Agency, so that he could begin preparing the steps necessary for the resale. Morison told him that he would be prepared,

8 Deputy Governor, 'Conversation with Mr Compton. Steel Bill', 5 Sept. 1952, File G1/125, Bank of England Archives.
9 C.F.C., 'Steel', 18 Sept. 1952, File G1/125, Bank of England Archives.

'somewhat unwillingly and as a public duty',[10] only under the following conditions: firstly, the Government must be really determined to carry denationalisation through; secondly, membership of the Agency Board had to exclude political appointees and be strictly a business board; thirdly, the idea of selling at market price had to be abandoned, with the basis being takeover price plus a loading; and fourthly, the Government had to accept that in the current market circumstances, it would be impossible to sell in a short period £300 million worth of steel securities, and therefore the Agency should be planned on the assumption that it would have to hold a large proportion of the capital of the companies in the form of low-interest-bearing debentures for some years to come.[11] Some days later Morison spoke to the Treasury and the Chancellor,[12] and then started work.

The following months saw a quickening of the political tempo. On 11 October 1952, the final day of the Conservative Party Conference at Scarborough, Churchill announced that the Bill to denationalise steel would be presented to Parliament in November. Four days later Hunter saw Sandys, who implied that it would be earlier rather than later in the month. Hunter assured Sandys that the BISF would give any help which the Minister required for this and subsequent debates, and submitted a memorandum on de-vesting. The main recommendation of the memorandum was that the preparatory work, and particularly the detailed work on individual companies, should be begun immediately. It seemed to Hunter that Sandys' attitude on this point had changed, '...and it was apparent that the importance of this is now realised in the Ministry. It is probable that, once the Bill is published and has had a Second Reading, satisfactory arrangements for preliminary studies and discussions with individual companies will be put in hand.'[13]

Two days after Churchill's speech to the Party Conference, Forbes called on Peacock and Phillimore at Barings to bring them up to date on steel and politics. First came the bad news: the Government had decided that it was politically undesirable to in-

10 C.F.C., 'Steel', 18 Sept. 1952, File G1/125, Bank of England Archives. Not the same memo as preceding reference.

11 Memo of meeting between Peacock, Phillimore and Forbes, 13 Oct. 1952, Partner's File 50A, f.106, Barings Archives.

12 C.F.C. 'Steel', 18 Sept. 1952, File G1/125, Bank of England Archives.

13 Chairman's Agenda, Executive Committee, 21 Oct. 1952, Box 011612, File July–Dec. 1952, BISF Papers.

clude any provision in the Bill to safeguard shareholders against renationalisation. Then more encouragingly, he suggested that perhaps the unions would not be intransigent. Lincoln Evans, the General Secretary of the Iron and Steel Trades Confederation, had said, privately, that until the Bill was introduced and debated, the unions would continue to declare their principled opposition, but that the Government should get on with it and put it on the Statute Book as soon as possible. Forbes hoped, as did the BISF, that Evans would agree in due course to serve on the proposed Iron and Steel Board.

Forbes told the bankers about Morison's meeting with the Governor, adding that he, Forbes, and Morison were as one in the sense of urgency. They both felt that if the operation was to be undertaken at all, it was important that six or eight of the leading companies should be resold as quickly as possible after the passage of the Bill; since the Government now estimated that this would be Easter 1953, that meant that the sale should take place between Easter and the end of July.[14] At least two months' hard work was frequently required to float an issue; Forbes and Morison were proposing that six to eight issues be floated within a three-month period beginning six months hence. It was no wonder that the BISF, the Bank and the Issuing Houses, as well as the future heads of the Board and the Agency (Forbes and Morison), were as one in urging the Government to authorise preparatory work to begin. The Government, however, feared that if such work were publicly authorised, it could jeopardise the interparty agreement over steel which they were still hoping to forge. Secret work, however, was a different matter, and decisions to begin would soon be taken. Meanwhile, with the debate on the White Paper, the political season for steel now began.

14 Memorandum of meeting between Peacock, Phillimore and Forbes, 13 Oct. 1952, Partner's File 50A, f.106, Barings Archives.

The Passing of the Bill

The Iron and Steel Bill made its Parliamentary journey in the usual spurts. The debate on the White Paper took place in the House of Commons on 23 October, followed by the First Reading in the House on 5 November, when the title of the Bill was read out by the Clerk and the date for the Second Reading named. The Second Reading took place later that month, followed by the Committee Stage, which began on 28 January 1953 and finished on 24 February. Then came the Report Stage, which took three days between 4 and 11 March, and finally, for the Commons, the Third Reading on 17 March. The following day the Bill went to the House of Lords for its First Reading there. The Second Reading took only one day, 31 March, the Committee Stage and the Report Stage following in late April. The Third Reading took only a short period on 5 May, and on 14 May 1953 it was announced in the Lords that the Iron and Steel Bill had received the Royal Assent.

The debate on 23 October 1952 was opened by the Minister of Supply, who moved that the House approve the Government's policy on iron and steel as set out in the White Paper. He pointed out that their two objectives were to restore the independence of the industry and to provide effective means for the supervision of the whole industry. Sandys went on to explain the position of the Holding and Realisation Agency, which would both hold and dispose. It would be financed by the Treasury, and any money it received over current needs would be paid into the Exchequer. The Agency would have very wide discretion in the planning and timing of the flotations, and the Government proposed to give the Agency power to effect amalgamations and modifications of companies' capital structures before offering them for sale. All of these operations, however, would have to be approved by the Treasury, who would be answerable to Parliament. When offered for sale, the price of the shares in the companies would be based on current values,[15] as would government stocks which might be exchanged for them. The time taken to sell the companies could run into years.

Sandys then turned to the powers and duties of the proposed Board. It would have powers relating to three specific subjects: development, supervision of prices and supplies of raw mate-

15 By 'values' Sandys presumably meant 'market prices'. This, however, would be tricky, since there was no market in steel shares.

rials. The proposals would implement the principles set out in the Election Manifesto, but there were also a number of features which Labour might consider were in part inspired by the Iron and Steel Board which they had set up in 1946 and by the T.U.C.'s Report of 1950. 'If so, we on this side of the House have no desire to be petty-minded. We are quite prepared to share with hon. Members opposite the credit for the paternity of this fine child.'[16]

Herbert Morrison, who had had overall responsibility for the 1945–51 Labour Government's nationalisation programme, took issue with Sandys' claim that nationalisation had meant increasing central control while denationalisation meant the opposite:

> I warn the right hon. Gentleman that if he is not careful, under the policy which he is pursuing, he will get more bureaucratic centralisation and interference under this Board — and remember it is the Board plus Steel House [the BISF] — plus the Agency — yes, plus the Ministry — I think that is about the lot — all of them floating about with powers of intervention of one sort or another.[17]

And who were going to become the new owners? 'We may get the denationalisation of iron and steel becoming the bankerisation of iron and steel.'[18]

Labour MP Frederick Mulley took up the theme that the Conservatives were imposing as much control as had Labour:

> ...quite frankly, this is not an issue between private enterprise and nationalisation. Control of some sort, or supervision, which is only the same word looked at from the opposite angle, is agreed between us....The issue is really a simple one: it is the issue of ownership. We say that if the industry is to be controlled, it can be effectively controlled in the national interest only if the ownership is in public hands.

The real issue was, who will get the profit?[19]

On the whole, the debate was bad-tempered. Conservative MPs naturally gave their support, and Donald Wade announced that he and his Liberal colleagues supported the White Paper,[20] but Labour MPs were against it to a man. Labour MP Jack Jones

16 505 H.C. Deb., 5s., cols 1274–91 (quotation in col.1291).
17 *Ibid*., col.1293.
18 *Ibid*., col.1304.
19 *Ibid*., cols.1312 (for quotation), 1313.
20 *Ibid*., col.1319.

wanted to know where the money — about £240 million — was going to come from?[21] George Strauss wound up for the Opposition. He underlined one irony of the situation, that in place of the 298 firms which the Iron and Steel Corporation controlled directly or indirectly, under the new dispensation supervision would extend over 2,400 firms,[22] the sort of comparison which tended to make Crookshank squirm. He then ended on a bitter note:

> How could we, with our unshakeable belief that the basic industries of our country must be under effective Parliamentary control and authority, and be run solely for the benefit of the nation, do nothing [sic] but regard this White Paper with unmitigated hostility?[23]

On 28 October 1952 the Cabinet approved the final draft of the Bill and authorised Sandys to arrange for its introduction at the opening of the new Session.[24] It was presented to the House of Commons on 5 November and published the following day. Hunter told the BISF Executive Council on 18 November that he had learned in confidence that the Debate on the Second Reading would be held on 26 and 27 November, but on 20 November the Cabinet decided that the debate would in fact be opened on 25 November by Sandys, while on the following day his Parliamentary Secretary, A.R.W. Low, would speak early and Macmillan would reply to the debate.[25]

The Second Reading was the most important stage of the Bill, when its main principles would be 'stated, attacked and vindicated'. The Bill itself would be discussed, without consideration of amendments, which would come during the Committee Stage.[26] When Sandys opened the debate, he concentrated on developing the arguments about supervision, implicitly attempting to answer the Opposition's taunts about the number of firms coming under the Board's control. He pointed out that steel

21 *Ibid* ., col.1329.
22 *Ibid*., col.1388.
23 *Ibid*., col.1391.
24 Cab. 90(52), Minute 4, 28 Oct. 1952, Cab.128/25.
25 Chairman's Agenda, Executive Committee, 18 Nov. 1952, Box 011612, File July–Dec. 1952, BISF Papers. Cab. 99(52), Minute 2, 20 Nov. 1952, Cab.128/25. The Transport Bill was given its First Reading on 5 Nov. as well.
26 Eric Taylor, *The House of Commons at Work* (London: The Macmillan Press Ltd., 1979, 9th ed., pb), p.94.

prices had effectively been under public supervision for twenty years. But Labour's Iron and Steel Corporation could not control the prices of any company which it did not own, and therefore to control through ownership would mean the nationalisation of hundreds of firms. Under the Conservative Bill, prices would be entrusted to a Board whose authority was defined in terms of a list of specific products, irrespective of the ownership of the companies who made them. He then moved to development, emphasising that the Bill provided two safeguards against the decisions of the proposed Board: the Board had to consult the trade association representing the branch of the industry concerned before exercising its powers, and there was a right of appeal from a decision of the Board to the Government.[27]

While Sandys emphasised the restraints on the power of the new organisations, George Strauss chose to emphasise what he saw as their lack of effective power to do much at all. The duties given to the new Board, he claimed, were the same as those of the old Corporation, but the old Corporation had all powers necessary over the ninety-two nationalised companies, while the new Board had 2,400 companies and virtually no powers at all. 'It is granted no authority whatsoever to ask any unit in the industry to do anything — to re-organise or anything else — which that unit does not want to do.'[28]

At 7 p.m. the debate on the Bill was interrupted for a debate on Kenya, but was resumed at 10 p.m. Labour MP George Chetwynd was just getting into his stride — 'Are the first offers for sale to be made to those who formerly held shares at the time of nationalisation, or shall we see the "Prudentialisation" of the industry?' — when proceedings came to an abrupt halt. Labour MP George Wigg indulged in what he termed 'Wiggery-Pokery': noticing that the benches held fewer than the forty Members required for a quorum, the Opposition demanded that the House be counted. Crossman noted in his Diary that 'as the Tories failed to muster, the debate on the Steel Bill collapsed.' The following day Crookshank announced that, in order to recover the time lost by the manoeuvre, the Government intended to devote 27 November to completing the Second Reading of the Bill as well as to the scheduled business. The result was an all-night sitting.[29]

27 507 H.C. Deb., 5s., cols.270–75.
28 *Ibid*., cols.283–88.
29 507 H.C. Deb., 5s., col.405. Morgan, ed., *Backbench Diary*, entry for 26

Nevertheless, the debate continued to be carried out in a much less emotional manner than had been the case with the White Paper, and the Government had a majority of thirty-six at the end of the debate, more than twice its majority in the House. This was somewhat surprising, since the Opposition might have been expected to draw encouragement from a recent Gallup Poll which revealed a marked drop in public support for denationalisation.[30] Perhaps the Opposition, however voluble in public, had privately decided that Sandys' Bill met most of their requirements, or if not, that the Minister was so keen to meet their wishes that he could yet be nudged in their direction. It must also have been crucial that the Iron and Steel Trades Confederation, and in particular Lincoln Evans, were not pushing the Parliamentary Labour Party to savage the Bill. Indeed, the evidence points in the opposite direction. On 2 January 1953 Evans, with five of his T.U.C. General Council colleagues, met with Sandys and some of his officials. Sandys stated that 'he hoped they would help him make the Bill as businesslike and workable as possible. Mr Lincoln Evans replied that the T.U.C. was willing to co-operate on that basis and endeavour to suggest improvements to the Bill.'[31]

The Second Reading had gone so well that Crookshank, the Leader of the House, reconsidered his earlier decision to impose a compulsory timetable for the remaining stages of the Bill. Instead, he decided to approach the Opposition to see if they could agree on a timetable.[32] They did, but not before the Opposition had extracted a promise of eight and one-half days for the Committee Stage of the Bill, which was then scheduled to begin on

Nov. 1952, p.184. Crossman thought the episode a bit schoolboyish. Lord Wigg, *George Wigg* (London: Michael Joseph, 1972), pp.166–67.

30 507 H.C. Deb., 5s., col.747. Question: 'Do you approve or disapprove of the Government's proposals about steel?' Replies:

	Oct. 1951	May 1952	10 Oct. 1952
Approve	43%	39%	35%
Disapprove	26%	34%	35%
Don't know	31%	27%	30%

Source: Chairman's Agenda, Executive Committee, 18 Nov. 1952, Box 011612, File July–Dec. 1952, BISF Papers.

31 T.228/357/131995. Evans was accompanied by Jack Tanner of the Engineers, J. Overs of the Blast Furnacemen, J. Gardner of the Foundry Workers, A.E. Tiffin of the Transport and General Workers and H.E. Mathews of the General and Municipal Workers, as well as Len Murray and V. Beak of the T.U.C. staff.

32 Cab. 101(52), Minute 4, 3 Dec. 1952, Cab.128/25.

28 January 1953.

While the Parliamentary proceedings were taking place, negotiations went on behind the scenes to try and square the various interest groups and to work out details of policy. Sandys and his Parliamentary Secretary, Low, met with Hunter and R.M. Shone, a Director of the BISF (and later, as Sir Robert Shone, a member of the Iron and Steel Board), on 9 December 1952. They discussed the provision of statistics, the relationship between the Board and the BISF and other topics.[33] He then met with the T.U.C. delegation on 2 January 1953, and there would be other contacts with interest groups. The Treasury and Ministry of Supply officials continued to work out details of possible amendments for proposal at the Committee Stage. The two major questions they considered were whether to impose a levy for development purposes and whether to merge some of the companies before selling them.

Early in 1952 it had been decided not to include in the Bill any provision for Government finance for the development of iron and steel companies once they had been denationalised. The Ministry of Supply, however, feared that the expansion of the industry would be held up for lack of funds, and tried by one means or another 'to find some method of doing the same thing by a backdoor.' In mid-January 1953 they came up with the idea of a compulsory levy on steel producers, with the funds being used to finance development by the guaranteeing of borrowing by companies, and asked the Treasury to consider including it in the Bill. The Treasury were not keen. For one thing, the decision had already been taken that the Board itself should not finance iron and steel development, and the giving of guarantees would fall into the same category. For another, the least creditworthy firms would be able to raise money for development, whereas part of the case for denationalising steel was that the financial mechanism of the market would squeeze out the least efficient. Finally, because the levy was proposed to help overcome the difficulties of financing development, which would remain while the threat to renationalise hung over the industry, it would be embarrassing for Ministers to have to include proposals which revealed that they thought this might be a real problem.[34]

33 'Note of a Meeting Held at Shell Mex House', 9 Dec. 1952, Box 01550, BISF Papers.

34 Figgures to Mynors, 26 Jan. 1953, for the quotation, and Minute by Figgures to Goldman, 21 Jan. 1953, both T.228/351/115995.

Nevertheless, it was a substantial proposal, and the Treasury decided that the Bank ought to be consulted. Mynors commented:

> When the operation has taken place, the Steel industry stands on its own feet among other industries: and if Government help is necessary in the provision of new capital, the best method will most probably prove to be one which is not confined to this particular industry. For what my judgement is worth, the Ministry's latest proposal would not conduce to selling Steel securities, quite apart from the confession of failure which any such addition to the Bill would now appear to be.
>
> ...I must say I am horrified that any responsible Department could appear to be so lacking in understanding of what this is all about. Their proposal, incidentally, is in terms of a guarantee of prior charges, suggesting that they do not even understand how an industry ought to be financed. This world of levies and subsidies in which they seem to live ....but I must not run on like this.[35]

Consequently, Figgures wrote to the Ministry of Supply that

> As I warned you on the telephone two or three days ago, we feel that your proposal that a levy should be imposed on steel products in order to provide a fund out of which to guarantee approved borrowings for development purposes is a non-starter. There are, in our view, a series of reasons for this. We do not believe it is right economically, we feel it would damage the prospects of sale and it appears to be thoroughly objectionable politically.[36]

This appears finally to have ended the Ministry's attempts.

The other question considered was whether to merge some of the companies before selling them, and if so, whether such mergers should take place before the new Board, which would carry some responsibility for the industry's development, was established and could be consulted. On the question of mergers, those within the industry might differ, but those outside tended to agree. Although one of the first acts of the new Conservative Government in November 1951 had been to issue a directive preventing Stephen Hardie, the then chairman of the Iron and Steel Corporation, from carrying out any company mergers and reconstructions, this did not mean that the Conservatives did not believe that they were needed. Rather, they felt that such mergers ought to be carried out by the industry itself. On the

35 Mynors to Figgures, 27 Jan. 1953, T.228/351/115400.
36 Figgures to Lindsell, 29 Jan. 1953, T.228/353/115400.

other hand, the Treasury grew to believe that this laissez-faire policy could lead to real embarrassment for the Government. As Figgures put it to his Permanent Secretary,

> Quite a number of the small companies ought to be merged with the main six or eight before sale. If this is done there is a reasonable prospect of recouping what those companies cost us and of avoiding a great deal of tiresome and long drawn out negotiation. If they are not merged with the six or eight main companies we must expect that these small ones will subsequently be sold at a loss.[37]

Thus, the Treasury supported H.G. Lindsell, the Under-Secretary of the Iron and Steel Division of the Ministry of Supply, when he raised the question as to whether the Ministry should be involved in 'encouraging' or forming mergers between the steel companies before they were returned to private ownership.[38]

Figgures drafted minutes arguing in principle for desirable mergers to be put in train, although he implied that difficulties were to be expected in specific cases. What drove Figgures was the tightness of the timetable. He began with the premise that many of the smaller or weaker companies ought to be merged with larger or stronger companies before their resale. Because the Bill could not now become law before mid-May 1953, the new Iron and Steel Board could hardly come into existence before June. Yet the core of Morison's scheme was to sell the equity of six or eight of the companies in the summer. If the mergers were desirable, but if they could not go through until the new Board had considered them, there was no chance that the timetable could be met. Therefore, either the larger companies would have to be sold without the smaller ones, thus risking a subsequent loss on the smaller ones, or the sale of the main companies would have to be delayed until the autumn.[39]

Figgures' draft was seen by Low, who attempted to strangle the babe at birth. He wrote to Sandys that

> It would be quite inconsistent with the arguments which we shall have to put forward in resisting the Opposition's Amendments relating to compulsory amalgamation that we should instigate or

37 Figgures Note to Gilbert, 10 Feb. 1953, T.228/353/131995.

38 Minute by Lindsell, 22 Jan. 1953, BT255/122, No.E3. Figgures to Lindsell, 3 Feb. 1953, T.228/353/115134.

39 Draft Minute, sent to Lindsell, 3 Feb. 1953, BT 255/122, No.E4A and Figgures' Note to Gilbert, 10 Feb. 1953, T.228/353/131995.

connive in rationalisation, re-grouping or mergers under the Corporation. Our argument is that desirable mergers or re-groupings will take place voluntarily between companies where there is a common interest. At this stage there can be nothing voluntary about amalgamations under the Corporation.

The substance of Low's Minute was sent to Lindsell, who sent it on to Figgures.[40]

Figgures, however, had taken the precaution of discussing matters with Morison, who fully supported and indeed encouraged Figgures in his arguments. Morison played his part by squaring Sir John Green, who had succeeded Hardie as chairman of the Iron and Steel Corporation and would in June 1953 be named a member of the Holding and Realisation Agency. Armed with Green's agreement, Figgures met with Sandys, who decided not to follow the advice of Low and turn the proposal down, but instead temporised. He would prefer, Sandys told Figgures, that Morison and Green should work out proposals for a list of mergers which should take place before the new Board was in place, and on the basis of the detailed proposals Ministers would later decide whether or not to go ahead with them.[41]

As far as Figgures was concerned, this did not solve the problem of the timetable, nor the political difficulty that many backbenchers, both Government and Opposition, were arguing that mergers desirable in the interests of the industry should be carried out before the companies were sold. He advised Gilbert that Morison should go ahead and plan, and then about 1 May 1953 ask Ministers whether the mergers should go ahead. Figgures then told Morison of Sandys' decision on 23 February 1953.[42] Morison, however, went about matters more directly. He discussed the situation with Forbes, the chairman-designate of the new Board, and they agreed that mergers not affecting the strategy of the industry should certainly go forward. Further (regardless of the fact that the official line was still that six to eight companies should be denationalised during the summer), Forbes

40 Low Minute to Sandys, 9 Feb. 1953, No.E6; Rowlands to Lindsell, 17 Feb. 1952, No.E7; and Lindsell to Figgures, 19 Feb. 1952, No.E8, all BT 255/122.

41 Figgures Minutes to Gilbert, 10 Feb. 1953 and 21 Feb. 1953, both T.228/353/131995.

42 Figgures Minute to Gilbert, 21 Feb. 1953, T.228/353/131995 and Figgures to Morrison, 24 Feb. 1953, BT 255/122, No.E9.

thought it unlikely that more than two issues could be made before the summer holidays and that such a small number of mergers would not prove too great an obstacle. The moment the Board was in being, the Agency should indicate the sort of mergers that they had in mind, and the Board could then respond immediately.[43]

Meanwhile, Morison and Green began, as Sandys had asked, to work on a list of proposals for mergers which might take place before the new Board was in place. Within a fortnight Morison was able to report encouraging talks with several companies. He told Mynors that among other possibilities, Dixon's Ironworks might go to Colvilles, and that Santon Mining and Templeborough Rolling Mills might both go to United Steel. However, difficulties then arose. Representatives of the Joint Iron Council came to Sandys and protested at the possibility of small concerns such as Dixon's being merged with bigger concerns such as Colvilles. Their fear was that Colvilles would then switch the furnaces currently making foundry iron to the production of basic iron for steel production.[44] Figgures let Morison know of the representations made to the Minister, but Morison was unrepentant:

> So far as Dixons is concerned, I am afraid our thoughts are likely to cause much trouble to the Joint Iron Council. I have discussed this Company with Green, and we are both satisfied that it would be desirable to tack it on to Colvilles if we can persuade that Company to take it. Green has already opened discussions with Colvilles on the point, and while it will not be easy to get their agreement we hope it may prove possible. The matter will then, I presume, go to the Minister of Supply. From the point of view of the Agency, Dixons is an unsaleable "dud" losing a packet of money, and, on financial grounds, it should be got rid of as soon as possible.

He went on to add that he and Green had met with about a dozen of the chairmen of the principal companies, when they had discussed the problems of the smaller companies in general terms; they would meet again when Morison and Green had some positive suggestions to offer.[45] Within a few days Green

43 Mynors' Note to Governor on talk with Morison, 25 Feb. 1953, File G1/125, Bank of England Archives.

44 Mynors' Note of a Conversation with Morison, 6 March 1953, File G1/125, Bank of England Archives. Lindsell to Figgures, 13 March 1953, T.228/353/131995.

45 Morison to Figgures, 19 March 1953, T.228/353/131995.

wrote to Sandys with a formal suggestion for the disposal of one of the smaller firms — the sale of the Glamorgan Hematite Iron Ore Company Limited to Guest Keen Baldwins Iron & Steel Co. Ltd — while the first disposal after the new Board was in place would be of another firm mentioned on 6 March, the sale in early August of Templeborough Rolling Mills to its former owners.[46]

While these discussions over policy and its practical application were taking place, the Parliamentary proceedings continued. There was a two-month break between the vote on the Second Reading, which signified the acceptance by the Commons of the principles of the Bill, and the Committee Stage which began on 28 January 1953, when attention would be focused on its details. The discussions were unusual in their productivity and good temper. After four of the eight and one-half days allocated had been used, Hunter reported to the BISF Executive Committee that

> The Debates themselves have been notable for the friendly atmosphere in which they are being conducted — the Opposition made it clear at the outset that they were whole-heartedly opposed to the Bill but they nevertheless intended to do all they could to see that the discussions were conducted in a spirit of goodwill.
>
> Clause 7 of the Bill has now been reached, and this still seems to be in line with the programme agreed between the Parties — it is, in fact, understood that the Opposition are a little concerned that they may not be able to make the discussion last the full 8 $^1/_2$ days for which they pressed the Government.[47]

This turned out to be the case. The Committee Stage was concluded on 24 February 1953 with Sandys and Strauss exchanging compliments. Only Crookshank, who had taken part in the negotiations with the Opposition over the timetable, was somewhat disgruntled, muttering into his Diary that 'I have always thought we had offered too much.'[48]

The Report Stage of the Bill was the last occasion on which amendments of substance could be made. The Government had promised to reconsider a number of points and therefore this

46 Green to Sandys, 24 March 1953, T.228/353/131995. *The Economist*, 8 Aug. 1953, p. 403.
47 Chairman's Agenda, Executive Committee, 17 Feb. 1953, Box 011614, File Chairman's Notes 1953, BISF Papers.
48 Crookshank Diary, 24 Feb. 1953, f.93, Ms d.361, Crookshank Papers, Bodleian Library.

stage took three days, beginning on 4 March, and continuing on 10 and 11 March.[49] And then, on 17 March 1953, the Bill had its Third Reading in the Commons. By this time, the Commons' consideration of the Bill had been very full indeed. Eighty-two amendments had been made — nine moved by the Opposition and twenty-nine by the Government to meet points made by the Opposition. The main changes can be summarised: the power of the Minister with regard to foreign supplies of iron ore were extended; the Board was to maintain 'close, full and continuous' consultation with the Government over relations with the European Coal and Steel Community; the Board was to consult widely with the representatives of producers, consumers, workers and merchants; the Board was to have future employment prospects as one of its concerns when considering future developments; and the Agency had a statutory duty to get a good price for the companies, while paying due regard to the efficiency of the industry and the need to leave it in good shape. In short, the details had been settled, and the Third Reading was primarily occupied with the re-statement of basic positions by Opposition and Government. The vote was taken, and the Government had a majority of thirty-three.[50]

The First Reading in the Lords took place on the following day, with the Second Reading two weeks later, on 31 March. The Marquess of Salisbury, the Leader of the Lords, introduced the Bill. He emphasised, as Sandys had done in his speech during the Third Reading in the Commons, that '. . . the Government are determined to get an adequate price for these shares and if, indeed . . . satisfactory offers are not forthcoming, the securities will not be sold.'[51] He pointed out that the Agency would not interfere with the day-to-day running of companies; however, it would be responsible for promoting efficient direction, primarily through its power to appoint boards of directors, and

49 Hubert Ashton (R.A. Butler's Parliamentary Private Secretary) wrote to Butler on 12 March 1953 that 'The Report Stage of the Iron and Steel Bill was safely concluded; the Minister appears to have done well and the Opposition co-operated as regards the time allotted.

. . . There seems little general interest in these Bills [Iron and Steel and Transport] either in the House or in the Country.' File G26, f.13, Butler Papers, Trinity College, Cambridge.

50 512 H.C. Deb., 5s., cols. 2086–94, 2202.

51 181 H.L. Deb., 5s., Col.362. In his speech Sandys had pointed out that this assurance was enshrined in Clause 19 of the Bill. 512 H.C. Deb., 5s., col.2192.

could merge or re-group companies before they were resold. The Agency would have to consult the Iron and Steel Board before re-grouping; if the Board refused, the Agency could appeal to the Treasury, who would be able to approve a scheme but only after laying a minute before Parliament and waiting two sitting weeks.[52] The procedure would clearly leave plenty of room for lobbying.

The Committee Stage took three days, 20 through 22 April, and a number of small amendments were made; the Report Stage then took place on 30 April. The Third Reading on 5 May 1953 consumed relatively little time,[53] during which the Government were magnanimous in victory and the Opposition spoke more in sorrow than in anger. As the Leader of the Opposition, Lord Jowitt, said in his closing remarks,

> Though I do not issue any threats, or anything of that sort,... it seems to us inevitable that if, in the whirligig of time, a Labour Government is once more restored to power, there will have to be another Bill dealing with the iron and steel industry.[54]

The Bill was then sent back to the Commons with the Lords' amendments.

On 14 May, the Commons having agreed to the amendments, the finished Bill returned to the Lords. In the end, it had only thirty-six clauses and two main points: (1) it dissolved the Iron and Steel Corporation and transferred the shares in the nationalised companies to an Iron and Steel Holding and Realisation Agency, whose sole function was to return them to private ownership; and (2) it established a supervisory Iron and Steel Board which would, in conjunction with the Ministry of Supply, exercise considerable powers with regard to prices, raw materials, capital investment and research. Later that day Mr Speaker and the Lord Chancellor announced the Royal Assent to the Bill,[55] and the Iron and Steel Act, 1953 was born.

52 181 H.L. Deb., 5s., cols.362-63.
53 Lord Swinton had written to Salisbury on 28 Feb. 1953 that 'I think I shall also get an agreed timetable on Steel, on which there is likely to be less talk [than on Transport].' File 174/6/1, Swinton Papers.
54 182 H.L. Deb., 5s., col.223.
55 182 H.L. Deb., 5s., cols.514 and 569.

The City Steams Ahead

While the civil servants and the politicians were re-drafting and debating the Bill, the bankers continued to plan for the actual denationalisation. The politicians wanted this kept as secret as possible, and Morison's appointment as adviser to the Treasury was not announced until late January 1953. Thereafter knowledge of the preparations spread more widely, but only when the Royal Assent had been given to the Bill was publicity encouraged rather than deplored.

Since mid-October 1952 Morison had been working on preliminary studies of companies which might be first in the queue for re-sale. He had been somewhat hampered: although the Bank and Treasury had agreed to this early in August, the Ministry of Supply held back until after the Second Reading. Consequently, Morison had information from the Iron and Steel Corporation but not from the companies themselves. He still brought along an 'ample' draft when on 17 November 1952 he called on Mynors at the Bank with his first Note, on the Lincolnshire company Stewarts & Lloyds (another on United Steel would follow a few days later). Morison wanted the Bank's opinion on whether it contained all the information required in an offer for sale: if so, he would produce a shortened version to be shown to Kindersley, Erskine and Phillimore — shortened because when bargaining between the Agency and the companies began, they might well be on the other side of the table, with, for example, Erskine acting for United Steel.

Morison had taken as a general working principle that the market value of the companies would be the price put on them at nationalisation in 1949 plus the profits retained since, and that this would be divided equally between equity and debt. On this basis, the value of the eight main companies was about £180 million, of which £90 million would have to be raised by selling ordinary shares and £90 million presumably would be held by the Agency, which would in due course fund it by selling quoted debt in the market. Morison had ascertained during a chat with the Chancellor that he would 'be content to be square on the operation *as a whole*, and thus the price could perhaps be raised for the stronger companies to compensate for the weaker.[56]

56 What Morison considered encouraging Hunter found disturbing: he believed that Morison's suggestion that the strong companies would have to be sold at an appreciably higher level in order to cover sales

What Morison now wanted was a chance to discuss with Erskine and Phillimore his ideas on the debt to equity ratios, yields and price/earnings ratios that should be offered when deciding the terms of the sales. He agreed with Mynors that the Bank's Working Party should be re-established as an umbrella for these talks.[57] Mynors asked Morison 'straight whether he would welcome the retention of Archie Forbes, assuming that he is going to the Steel Board. He said that Forbes was of course most knowledgeable but not always entirely discreet.'[58] As it turned out, Forbes was abroad. The Bank added Sir Thomas Frazer to the Working Party, an actuary who was a Director of North British and Mercantile Insurance Co. Ltd, and since 1936 a member of the Government's Capital Issues Committee, which determined which issues could be floated on the Stock Exchange.

This reconstituted Working Party met with Morison at the Bank on 3 December 1952. They had already received copies of shortened versions of Morison's Notes on Stewarts & Lloyds and United Steel, as well as a list of points on which Morison wanted advice. There was considerable discussion about the preferred debt to equity ratio, with the point being emphasised that there was considerable appetite for fixed-interest securities.[59] The conclusion was that a ratio of debt to equity of less than 50% was probably impracticable. Another conclusion reflected a change in the assessment about how best to begin the resale. The prevalent opinion had always been that as large a group of the larger companies as possible should be sold at the outset. Now, the Working Party felt that the best method of testing the market would be to offer the whole of the equity of one of the principal companies, with great care taken to minimise the risk of

of the weaker companies at a loss was a considerable departure from the previously discussed basis. Memorandum of meeting, 11 March 1953, Partner's File 50A, f.128, Barings Archives.

57 H.C.B.M., 'Steel', 17 Nov. 1952 for the quotations and Mynors' Note for the Governor, 'Steel Working Party', 25 Nov. 1952, both File G1/125, Bank of England Archives.

58 H.C.B.M., 'Steel', 17 Nov. 1952, File G1/125, Bank of England Archives.

59 This was especially important to insurance companies, since the policies they wrote were predominantly to provide a specified sum of money at a specified date in the future. Thus they preferred fixed-interest securities, i.e. debt, rather than equities, which might or might not pay the assumed stream of dividends. Thanks to Jeremy Wormell for the suggestion.

failure.[60] But this was the opinion only of those attending the meeting, and certainly others, including some politicians, continued to believe that a group of companies should constitute the initial offer. Nevertheless, when the time came, it would be a single concern which was offered.

Morison refrained from opening negotiations with the steel companies themselves until the Chancellor announced on 23 January 1953 that he had appointed him as his main adviser on de-vesting. Thereafter he moved quickly. He saw Hunter near the end of the month and told him that he hoped to start talks with the companies at an early date. Hunter alerted the companies, and United Steel contacted Morgan Grenfell.[61] Walter Benton Jones, the chairman, and another Board member met the bankers in early March,[62] and thereafter Morgans, and in particular Erskine and his colleague Kenneth Barrington, worked to ensure that United Steel would be prepared, whenever its turn came to be floated.

By this time the Committee Stage of the Bill had been completed, and the good humour which had prevailed during the debates considerably encouraged the BISF and the bankers. At a meeting at Barings on 11 March 1953, Hunter mentioned the marked improvement in the political atmosphere surrounding steel. For one thing, Lincoln Evans had been telling his trade union colleagues that whatever their views, their duty was to co-operate with the government of the day. But equally important from Hunter's point of view, 'the Socialist Party's opposition to the measure also appeared much less positive than it had been.' During the recent debates on the Committee Stage, 'Strauss did not repeat the threat' to renationalise but took the line that his Party did not believe that the new set-up would work, and if it did not, they might have to take the industry back into public ownership.[63] This conditional approach was much less unnerv-

60 Minutes of Working Party, 3 Dec. 1952, Partner's File 50A, f.116, Barings Archives. This would almost certainly be the way an Accepting House would suggest doing the job now.

61 Chairman's Agenda, Executive Committee, 17 Feb. 1953, Box 011614, File Notes 1953, BISF Papers.

62 Peddie to Erskine, 27 Feb. 1953, and Erskine to Peddie, 2 March 1953, both File United Steel Companies 1953 Offer for Sale of £14m £1 Ordinary Shares (hereafter File USC 1953), Box P377, MGP.

63 Memorandum of meeting, 11 March 1953, Partner's File 50A, f.128, Barings Archives.

ing than his positive threat to renationalise, made on 12 November 1951.

The Treasury waited until the Bill had gone to the Lords; then on 18 March the Chancellor and the Governor agreed that the Bank and the City 'had better get on with this market side without bothering them [the Chancellor and the Treasury] too much.'[64] Morison had spoken to Erskine, who no longer believed that the Bank should make the issue. He thought instead that there should be a syndicate of the six or seven big Issuing Houses, with the main work on each sale allotted to whichever House usually advised the company concerned.[65] The Bank set to work accordingly. The Governor spoke to Peacock and to the Hon. R.H.V. Smith, son of Lord Bicester, and himself a director of Morgan Grenfell. They accepted the need for a Consortium, but they thought that the offers should be made by the Agency, rather than the Agency selling the shares to the Consortium who would then re-sell them to the public. They argued that the Bank should be the receiving bankers in form, even if the clearing banks did the work. They also thought that the first offer should come after the Iron and Steel Board was set up so that its membership would be known.[66]

All three agreed that the crucial element was the attitude of the insurance companies. The Governor suggested that he see the companies and try to arrange for them to set up a working party which would work with one to be set up by the Consortium of Issuing Houses. 'After the initial opening, he would leave the two groups to work together in order to minimise the risk of the whole operation appearing to be a City huddle under the chairmanship of the Bank.'[67]

Meanwhile, he recommended to the Chancellor that Morison have initial discussions with a group of Issuing Houses, who would act for the Agency in marketing the securities to the public; this group would then use the services of the leading brokers.

64 Governor's Note, 18 March 1953, File G1/125, Bank of England Archives.

65 H.C.B. Mynors, 'Memorandum', 17 March 1953, File G1/125, Bank of England Archives.

66 This was inevitable in any case, since according to Clauses 1(1) and 2(1) of the Act, the Agency could not act until after the appointed day, on which the Board would also come into existence. The date would be 13 July 1953.

67 H.C.B.M., 'Steel', 20 March 1953, File G1/125, Bank of England Archives.

He further recommended that the group should begin taking soundings with the large institutional investors. The Governor admitted that there was some risk that the preparations would give rise to talk, but he hoped that the Chancellor would agree 'in thinking that this risk must be taken now that the Bill has reached a fairly advanced stage.' He also emphasised that it was most important that the operation should be handled, as far as possible, on ordinary market lines by those most experienced. If the Chancellor and the Minister of Supply agreed he would proceed to get a group formed and encourage them to get in touch with Morison on the one hand and leading institutional investors on the other.[68]

Both Butler and Sandys agreed that the Governor should form a syndicate, since if any issues were going to be made during the summer, it would have to be set up well before the Royal Assent was given. On the other hand, Sandys wrote,

> I am most anxious that we should run no risk of disturbing the calm political atmosphere, which we have been at such pains to create. [69] There would in my opinion be a serious danger of this, if it were to appear that the Government was itself approaching prospective buyers before the Bill reached the Statute Book. But I regard it as quite a different matter for the City to begin making preparations on its own account.
>
> To sum up, my view is that it is very necessary for the Issuing Houses to get moving as quickly as possible, and that there is no objection to their making discreet soundings in the right quarters. On the other hand, I consider it essential that, in their contacts with outside interests, the group should make it clear that... they are not as yet authorised to enter into any binding commitments on behalf of the future Agency. In fact, the operation should, as far as possible, be represented at this stage as a City rather than a Government initiative.[70]

While awaiting the Minister's response, the Governor got on with his preparations. He spoke to Mr Ferguson of the British Insurance Association, who strongly opposed Cobbold's idea of

68 Cobbold to Butler, 25 March 1953, File G1/125, Bank of England Archives.

69 Or as Figgures put it, 'The Minister of Supply's great achievement is that he has made steel denationalisation into a bore but we cannot count on this continuing.' Figgures to Gilbert and Couzens, 14 April 1953, T.228/353/115134.

70 Sandys to Butler, 1 April and handed by Butler, who concurred, to Cobbold, 2 April 1953, File G1/125, Bank of England Archives.

forming a working party amongst insurance company general managers. He suggested, rather, that the Governor tell the chairmen of the Prudential Assurance and the Pearl Assurance what was afoot, and he himself would have a private talk with some of the general managers. Ferguson strongly supported the view that the operation should be done as much as possible through market channels, and he thought that a more definite approach to the insurance companies should be made individually through the normal market machinery, i.e. by the brokers. Cobbold subsequently saw the two chairmen, who confirmed that the normal market approach was the right one. The Governor therefore gave up the idea of an insurers' working party.

He then turned to the Issuing Houses, seeing individually Olaf Hambro, Hugh Kindersley, Anthony de Rothschild, Helmut Schroder and Alfred Wagg, all of whom agreed in principle to co-operate. This agreement, however, masked strong disapproval on the part of the Rothschild partners and Ashburton of Barings. At a meeting at the Bank on 8 April attended by representatives of eight Houses — Evelyn Baring characterised them as the usual six (Morgans, Barings, Rothschilds, Schroders, Hambros and Lazards) plus Helbert, Wagg and Robert Benson, Lonsdale — Anthony de Rothschild expressed his (and Ashburton's) view that the way the Governor was going about the business was quite wrong.[71] On the following day, at a more formal meeting of representatives of the Houses plus Bank Directors, de Rothschild

> expressed doubts whether the sale of the steel shares could be conceived as an operation to be handled through ordinary market channels and wondered whether some quite different method would not be preferable: he felt that the more one looked at it, the more difficult the operation appeared.' The Governor was solicitous and offered to arrange further meetings to discuss the matter, but he also held to his view that 'the sale of shares in iron and steel companies was a Government policy decision and it had to be made a success; there was need for the full City machinery to be used for this purpose, if only in the interest of the City itself.[72]

71 Governor's Note, 27 March 1953, File G1/125, Bank of England Archives. E.B.B. to Ashburton, 8 April 1953, Partner's File 29A, f.14, Barings Archives.

72 'Confidential Minutes of meeting held 9 April', 13 April 1953, File Steel Industry 1953 *et seq*. Denationalisation. General. Documents and Memoranda (hereafter 1953 Denationalisation General, Docs/Memos), Box 369, MGP.

This was an argument which David Colville of Rothschilds found unacceptable:

> This whole matter is far too political. Nationalisation was a spiteful act, carried through without a mandate, and denationalisation is being effected to redeem a party pledge. The City is much best kept out of politics.
>
> What the Bank of England is in fact doing is to use its influence in the City, acting as the nationalised agent of a Government controlled by a political party, to obtain the support of private enterprise to help implement the election promises of that party. In effect, private enterprise is being suborned by a nationalised institution for political ends and the Bank is, moreover, seeking to throw the whole political onus and possible loss on the City, without taking either risk or responsibility.[73]

His objections, however, seem to have had little impact. Perhaps the other Houses feared a conflict with the Governor, the acknowledged leader of the City. More likely they agreed with him that it was important, whether on political, economic or philosophical grounds, that the steel companies be returned to private ownership. The concerns of the bankers were not, strictly speaking, party political: no-one was minuted as stating that denationalisation had to be carried through either because it was Conservative Party policy or because it would undermine the policy of the Labour Party. However, the Labour Party was always referred to as the 'Socialist party', and implicit in the discussions was the conviction that a limit should be put on the growing tendency of governments to interfere in industry and business. After a further meeting of the Issuing Houses on 13 April, Phillimore had a third explanation for the lack of support for Colville's objections: 'In general, I got the impression that all of the Houses present, with the exception of Rothschilds and us, were now in full cry after their respective [steel company] businesses.'[74] In short, political predilections and self-interest combined to ensure that the City would play the part assigned

73 David Colville, Private and Confidential Memorandum on Steel sent to Morgan Grenfell [and presumably the other Issuing Houses], 14 April 1953, File 1953 Denationalisation General Docs/Memos, Box 369, MGP. Colville was the first non-Rothschild to be a director of a Rothschild bank.

74 Phillimore to Ashburton, 13 April 1953, Partner's File 29A, ff.19–20, Barings Archives. Phillimore of course did not want Barings left out, adding that 'We must somehow find out, without delay, whether Ellis Hunter and Latham propose to come to us or whether they are

to them.

The Governor had imposed his view that the business should go through the normal City channels, with the Issuing Houses rather than the Bank taking the responsibility, and he now disbanded his Working Party. In its place it was proposed to set up a small Working Committee of representatives of the Houses, and a meeting was called at the Bank for 9 April 1953 to set up this machinery formally. At the meeting it was decided that the Houses themselves would establish the Committee, but the Governor set out the general principles on which they should all work. First of all, the operation had to be made a success, in the interests of the City itself. Secondly, a broad framework had to be established, rather than a 'piecemeal attack company by company sponsored by individual Houses.' Those who had previously advised the major steel companies should together devise such a framework, in consultation with Morison and potential investors; these discussions should remain confidential, in particular until the Bill became law. Approaches to the institutions should be made individually and through the ordinary channels, and it was agreed that the normal Stock Exchange machinery should be used.[75] In short, the goal was, in an abnormal operation, to act as normally as possible — the approach least likely to scare off the institutions.

On 13 April 1953 representatives of the Issuing Houses met at Morgan Grenfell to set up their Working Committee and begin their own preparations for the sale. The first order of business was for Erskine to let the others into the secret of the Bank's Working Party. He noted that Morison had been a member, and on the basis of their advice as well as that of other City interests, Morison had decided the lines upon which the sales would take place. The first stage would be the sale of the equity of several large, formerly publicly-quoted, companies[76] and the sale of former wholly-owned subsidiaries. The second stage

making other arrangements; and we are just now discussing how to do this without actually asking them point-blank.'

75 E.B.B. to Ashburton, 8 April 1953, Partner's File 29A, f.14, Barings Archives. Confidential Memorandum on Iron and Steel, 13 April 1953, File 1953 Denationalisation, Box 369, MGP.

76 These were Stewarts & Lloyds, United Steel, John Summers, Colvilles, Dorman, Long, Lancashire Steel and possibly Whitehead. The aggregate market value of their equity would have to be kept down by 'gearing', i.e. taking on debt, in order to reduce the amount of money required from the public. Richard Thomas &Baldwin and the

would cover the equity of the smaller companies and the debt and preference shares of the large companies. The large public companies would be recapitalised using estimated market values with about half being equity. The other half would ideally be in the form of redeemable preference shares so as to hold the issue of debt available for the future.[77] The prices which the Agency would set for the shares of the large public companies, Morison expected, would reflect 'the take-over price with a loading to take into account some part of the retained profits during the period of State ownership and the expenses of the operation, such loading to provide a cushion against losses on subsequent transactions.' Opinions on this had wavered back and forth, but the principle was now decided. Finally, the others learned that the sum involved in the sale of the first stage public companies would be of the order of £80 million to £90 million.[78]

Erskine's explanation was followed by discussion, from which emerged general agreement on certain points. First of all, the Realisation Agency should appear as vendor in the prospectuses.[79]

Steel Company of Wales would have to be kept out of the first stage because of their very heavy burden of debt.

77 That is, they could in future borrow by means of debentures and loan stock if new capital was required. Morison also thought that the nominal capital of the large public companies should be increased by the capitalisation of reserves in order not to have too high a rate of dividend on the ordinary shares. Too, the ordinary dividend might have to amount to something like 50% of what might be regarded as the net maintainable profits. Appendix A, Minutes of Meeting of Issuing Houses Steel Committee, 13 April 1953, File 1953 Denationalisation, Box 369, MGP. In their Report of 13 May 1953, the Steel Sub-Committee wrote that 'As a general guide we consider that on the basis of yields ruling to-day the appropriate yield basis for steel shares, provided the "gearing" and cover are reasonable, would be:-
(a) Ordinary Shares — 6 3/4% to 7%
(b) Redeemable Preference Shares — 5 1/2% to 6%.
There will obviously be differences between the various companies above and below these yields.' 'Report by the Sub-Committee on Steel Appointed by the Eight Issuing Houses', 13 May 1953, p.6, File 1953 Denationalisation, Box 369, MGP.

78 Appendix A, Minutes of Meeting of Issuing Houses Steel Committee, 13 April 1953, File 1953 Denationalisation, Box 369, MGP.

79 'It was understood that the Governor was not in favour of the Bank of England actually making the Offer for Sale, e.g. as for example it does for a colonial government stock. Various views were expressed as to the desirability of asking the Governor to reconsider the question and

Secondly, it would be a complex task to exchange Government securities for steel shares, and it was probably essential that the Bank act as receiving banker (to which it had reluctantly agreed). Thirdly, the Houses represented at the meeting would form a Consortium to advise both the Agency itself and the directors of the individual companies whether the proposed terms of sale were fair to all parties, as well as being acceptable to the Consortium as the basis for arranging the underwriting. At the same time, the individual Houses would continue to advise their clients and negotiate with the Agency on their behalf. Here, the Committee's responsibility would be limited to ensuring that this advice was based on generally accepted principles and that all the companies received the same treatment. Finally, they decided to set up a small sub-committee to consider details of procedure; the members were J. Backhouse of Schroders, J.H. Hambro of Hambros, P. Horsfall of Lazards, A. Russell of Helbert, Wagg and Erskine of Morgan Grenfell.[80]

The Sub-Committee had its first meeting immediately after the end of the main Committee's meeting. Erskine became chairman, and Barrington and K.F. Chadwick of Morgan Grenfell became Joint Secretaries. The Sub-Committee plunged immediately into detailed discussion of a Memorandum Erskine had drawn up in March, which was based on the earlier discussions both of the bankers' Working Committee and the Bank's Working Party, and which set out the procedural suggestions made by those earlier groups. It was also assumed that firm subscriptions should be obtained before the public were invited to subscribe — that is, that agreements would be made covering the purchase of 75% of the total issue. This would need to be stated in the offer for sale documents. 'This it is hoped would encourage outside applications resulting in the allotments to the institutions being substantially scaled down which would be very helpful vis-a-vis

particularly whether such a procedure would offer some additional security against the political risk.' Minutes of Meeting of Issuing Houses Steel Committee, 13 April 1953, File 1953 Denationalisation, Box 369, MGP.

80 *Ibid*. Phillimore reported to Ashburton on 13 April 1953 that Barings, Rothschild and Bensons were not represented on the Sub-Committee, adding that 'it is no doubt the fact that we have no particular associates among the leading Steel Companies that accounts for our being left out of the Sub-Committee.' Partner's File 29A, ff.19–20, Barings Archives.

subsequent offers.'[81]

The probable response of the institutional investors continued in the forefront of discussion. There were two aspects to the problem: would they invest in the shares of denationalised steel companies at all, and if so, to what extent could they commit themselves? It was hoped that individual discussions would convince the insurance companies in particular that they should be heavy holders. The other factor was simply the question of whether the money was available. The size of the whole operation was estimated at £300 million. Of this, about £55 million concerned former private firms which might be resold to their previous owners without any call on the market's resources. The total capital of the eight large formerly publicly-quoted steel companies which were most likely to figure in the operation in the immediate future was calculated to be £196.9 million, of which equity might be about £102.3 million. (That left about £100 million for others, such as the Steel Company of Wales, which would be dealt with at a later stage.) The long-term investment institutions' resources were divided as to insurance companies about £3,000 million, pension funds about £750 million and investment trusts about £500 million. Of the insurance company portfolios, £320 million were in equities. The assumption was made that 10% of pension fund assets or £75 million, and 75%, or £375 million, of investment trust assets were equities. The total came to £770 million. The final assumption was then made that 10%, 10% and 5% respectively of these equity exposures might reasonably be expected to be invested in the steel industry, making a total of £54.5 million.[82] When this was compared even with the £80 to £90 million which Morison estimated was required, the gap between the two was daunting. Consequently the first decision which the Sub-Committee took was that the sales must be public offers directed at the private investor as well as the institutions, a decision which would lead to high costs for advertising, with large sums falling into the laps of numerous regional newspaper proprietors. After further discussion, the Sub-Committee adjourned until 16 April.[83]

81 Appendix C, Minutes of Meeting of Sub-Committee on Steel, 13 April 1953, File 1953 Denationalisation, Box 369, MGP.

82 N.n., 'Steel', 9 April 1953, Partner's File 29A, ff. 16–17, Barings Archives.

83 Minutes of Meeting of Sub-Committee on Steel, 13 April 1953, File 1953 Denationalisation, Box 369, MGP.

The Governor decided to warn the Chancellor and the Minister of Supply of the difficulties envisaged by the Sub-Committee, especially the possible lack of funds. After warning them — as he had done months before — that 'one or two sound judges with many years of experience have told me that it looks like the most difficult task with which they have ever been faced', he added sombrely that 'I am by no means confident that, with the political risk involved and the dearth of new private savings, the offer of Steel securities will produce any substantial response from private investors, and there are definite limits to the lengths to which, with the best will in the world, institutional investors could properly go.'[84] On the other hand, the Governor received more encouraging news on 1 May, when he spoke to Ferguson of the British Insurance Association. Ferguson had spoken to a number of his London colleagues, as well as to some of the Scottish managers in Glasgow, and 'he appeared to think that the climate was fairly good.'[85]

While the procedural details were being worked out, there was also the membership of the Agency to be considered. Gilbert of the Treasury wanted Sir John Green, the current head of the Iron and Steel Corporation, to be a member of the Agency, and he asked Morison to 'square the Chairman-designate of the Board' [Forbes]. Morison knew that if he asked Forbes directly he would 'get a dusty answer.' He decided to hold back for a while, on the grounds that if he waited Forbes would accept it, and at any rate, he would not be able to make it a condition of his own appointment to the Board that Green not be a member of the Agency. The Governor persuaded Sir Oliver Franks, the former British Ambassador to the U.S. and currently the Deputy Chairman of Lloyds Bank, to become a member of the Agency, and A.C. Bull, the former Principal of the Discount Office of the Bank of England, was also to be a member. On 28 April the Governor and Morison decided that if these names went forward, along with that of Sir Thomas Chadwick, the former Accountant of the Treasury, they need not look any further for the moment.[86]

84 Cobbold to Chancellor, 14 April 1953, copy to Minister of Supply, File G1/125, Bank of England Archives.
85 Governor's Note, 1 May 1953, File G1/125, Bank of England Archives.
86 H.C.B.M., 'Steel: Note of a Conversation with Sir John Morison: 22 April 1953', for quotations, and Governor's Note, 1 May 1953, both Bank of England Archives.

Meanwhile, the Sub-Committee continued to thrash out the details. They met with Bank officials on 30 April to discuss the form of the offer and of the publication of prices at which Government Stock could be tendered in exchange for shares. They also wanted to sort out the approach to the Clearing Banks. Once agreement had been reached, the Sub-Committee finished its Draft Report. This was then sent to each of the Issuing Houses and their comments invited. Once these had been received and changes made in the draft, the final Report, dated 13 May 1953, was distributed to the Issuing Houses, the solicitors, Slaughter and May and Linklaters & Paines, the accountants Peat, Marwick, Mitchell, the Bank of England and Morison.[87]

The Report incorporated many of the points which had been discussed over the months, but it also included a condition which was to disrupt the hoped-for schedule. The Sub-Committee accepted that the sale of the equity of the large, publicly-quoted companies should form the first stage of the operation, but they insisted that it 'must proceed concurrently with the resale of former privately-owned companies, e.g. English Steel and Guest Keen Baldwins.' This had come up in earlier talks, but it had not apparently been a major point of discussion. Now the Sub-Committee stated firmly that the Consortium must not be expected to commit themselves to the first offer for sale until it was publicly known that negotiations for the reacquisition of the major privately-owned companies were in an advanced stage. No reason was given. Morison had emphasised the difficulties inherent in carrying on simultaneous intricate negotiations for both quoted and unquoted sales, but the Sub-Committee nevertheless thought it essential, adding in the Report that 'we accept the fact that insistence upon this point may delay the date of the first public Offer.' It was indeed to do so.

The Report then considered whether in the first instance the whole or only part of a company's equity should be sold. It was true that selling, say, 60% would reduce the pressure on the market, as well as avoiding the danger that after one or two successful issues there would be insufficient 'risk bearing' cap-

87 'Points for discussion with the Bank of England at the Meeting to be held on 30th April', and 'Draft Report by the Sub-Committee on Steel', 26 April 1953, both File 1953 Denationalisation, Box 369, MGP. Erskine to Colville, 6 May 1953; Jack Hambro to Erskine, 6 May 1953; Evelyn Baring to Erskine, 7 May 1953; and Rex Benson to Erskine, 8 May 1953, all File 1953 Denationalisation, General, Box 369, MGP.

ital available to finish the operation. However, the arguments in favour of selling 100% of the equity were compelling. For one thing, offering only 60% would be a public admission of difficulty; further, steel companies would hardly wish to have, indefinitely, the Government as a 40% shareholder, with the implied threat of interference; and finally, the market price of the first tranche of shares might be adversely affected by having a further block overhanging the market. 'On balance therefore we favour planning on the basis of selling the whole of the equity unless it is proved in the course of negotiations that this is too ambitious.'

With regard to the capital structures of the large publicly-quoted companies, the Sub-Committee accepted Morison's plans as a general guide. However, they thought that it was vital to ensure that they were soundly based as they went back into the private sector, 'even if such a course requires the provision of a larger amount of money by the public for the equity.' Therefore, a conservative approach should be taken to gearing, i.e. the amount of debt with which the companies were sold. The Sub-Committee reserved comment on the Agency's expectation of a value above the takeover price, preferring to wait until the schemes for individual companies were prepared. As for the order in which the companies should be brought to market, 'the figure for the equity of Stewarts & Lloyds, which has a high earning capacity, will no doubt be a large one, but that company has probably the greatest investment appeal and, notwithstanding the size, we favour it as the subject of the first Offer for Sale.'

The Report then turned to financial relations between, and financial recompense for, the Issuing Houses. 'We fully concur with the Governor's request that care should be taken to avoid any accusation that "the City" is making undue profit and we are of the opinion that the percentage spread chargeable to the Agency should be kept as low as possible.'[88] The profit to the

88 The spread was to be based on the following: (1) sub-underwriting or commitment commission should be on a generous basis in the light of prevailing market conditions; (2) the need for subscriptions by the private investor meant that allotment brokerage needed to be on a generous basis, especially since agents would have heavy expenses in handling the tendering of Government securities; (3) overriding commission should be possibly up to $1/8$% on the money, expressed in the form of pence per share, and would be divided amongst the consortium of brokers who would not receive allotment brokerage on

Houses should be of the order of 1/4% on the money,[89] expressed as pence per share, to be divided equally amongst the members of the Consortium. In addition, those banks which advise companies on the reorganisation of their capital structure should receive an additional fee which they would retain.

The Sub-Committee agreed that the form the offers for sale would take should conform to a common pattern. The Bank of England would appear as the receiving bank. Shares could be paid for either by cash in instalments, or by tendering any permissible Government security, and as the operation would be in essence a conversion offer (because of the option of tendering Government stock) as much as a cash offer, the subscription lists should be kept open for between seven and fourteen days. One problem which remained to be solved was whether reference should be made in the prospectuses to the renationalisation threats made by the Labour Party: non-disclosure of a material fact in a prospectus was illegal. The Sub-Committee's opinion was that it was a matter primarily for the Agency and the companies.

The Agency emphasised that success would require the cooperation of a wide range of City bodies. The Sub-Committee had decided that Erskine and Horsfall would liaise with Mynors and Beale of the Bank, while the Governor of the Bank would approach the Chairman of the Committee of London Clearing Banks. Negotiations would have to be opened with the Chairman of the Council of The Stock Exchange, since changes in some of the rules would have to be arranged. To handle the various operations, the Sub-Committee thought the Houses should form a consortium of London brokers,[90] although the broker who usu-

guaranteed applications. 'Report by the Sub-Committee on Steel...', 13 May 1953, p.7, File 1953 Denationalisation, Box 369, MGP.

89 Evelyn Baring had written to Erskine on 7 May 1953 that 'We assume that the Sub-Committee is satisfied that it will be possible to get firm subscriptions for 100% of the Stock offered from Institutional investors. We assume that the eight Houses will not be planning to take the Stock themselves but only to secure subscriptions. If this be so, it seems to us that the 1/4% commission suggested as the profit for the Houses may be open to the kind of criticism which the Governor of the Bank of England is so anxious to avoid.' File 1953 Denationalisation General, Box 369, MGP.

90 Cazenove Akroyds & Greenwood & Co., W. Greenwell & Co., Hoare & Co., Panmure Gordon & Co., Rowe & Pitman and Joseph Sebag & Co.

ally acted for a company would take the lead in any given case. For legal advice, all the members of the Consortium would use either Linklaters & Paines or Slaughter and May, and they would act jointly for the Consortium; the Agency would be advised by Freshfields, the Bank of England's solicitors. The three firms together would draft the skeleton documents. If the advice of an independent accountant was needed, they would consult Sir Harold Howitt of Peat, Marwick, Mitchell.

Finally, the Report considered how best to approach the object of all of this planning, the investing community. The Sub-Committee were as one in their conviction that the offers for sale would have to reach the widest possible number of investors, which meant advertising widely. When the time came for the individual Offers themselves, the approach to the institutions would as usual be made by the brokers. However, in the immediate future personal contact should be made with each of the leading insurance companies, who had repeatedly expressed distinct doubt about the whole operation, and with the leading groups of investment trusts, in order to explain the plan and engage their support. Pension funds, however, were probably best taken care of by the consortium of brokers in the usual way.

One question which had exercised all those involved in the planning since the beginning was what to do about former steel company shareholders. Agreement had been easily reached that these holders of 'pink' forms should have priority in allocation of the shares, but what should be the 'record date'? The Sub-Committee had finally decided that it should be the vesting date: that would provide the majority of those who had held steel securities during the years 1949 to 1951 with a pink form, and would certainly ensure that all those dispossessed on vesting day received one. It was also, according to the Bank, the only practicable date.

But these former shareholders, as well as possible new ones, had to be convinced that it would be a wise move to invest in the denationalised companies. In the circumstances, the Sub-Committee attached 'considerable importance to ensuring adequate arrangement for liaison with the Press. It is desirable that, well in advance of the first Offer for Sale, the Press should be provided with adequate information regarding the operation as a whole so that public interest may be aroused by well-informed and realistic press comment.' They hoped that Morison would arrange a press conference shortly after the Agency had been formed.[91]

This Report, then, suggested how best to carry out the policy so painfully put together by the Government. The Committee drew up a schedule (or battle plan) setting out who had to do what and when, so that once the Bill became law, the operation could begin. Whether fortuitously or not, the day the Report was distributed — 14 May 1953 — was the day the Iron and Steel Bill, 1953 received the Royal Assent. The previous ten months had seen, both in the East and West End of town, a great deal of effort devoted to legalising and organising the resale of the nationalised steel companies to the private sector. Now came the big test: would anyone want to buy them?

91 'Report by the Sub-Committe on Steel' 13 May 1953, File 1953 Denationalisation, Box 369, MGP.

4 SELLING THE SHARES

The period from May to October 1953 saw the climax of the previous two years' preparations: the bringing of the first steel company to market. The plan had always been to sell off a group of companies during the summer of 1953, but conditions imposed by the institutions prevented this. Even after it was accepted that the earliest a sale could take place was October, it still took some weeks to decide that the United Steel Companies should be the first to go. Initially, the operation was a resounding success; but the price in the aftermarket did not hold up, and by the following February it was branded a failure.

During the actual selling operation the politicians were barely involved, and this part of the story takes place almost wholly in the City. In the first section the focus is on the general problems which remained to be solved, such as whether a private company ought to be resold before the shares of the first publicly-quoted company were issued, and what was the best way to approach the institutions. The second section concentrates on the issuing of the shares of the United Steel Companies, while the third section assesses the short-term success as well as the medium-term failure of the operation.

Deciding What to Sell, and When

The Report of the Issuing Houses Committee, which had been distributed on 14 May 1953, remained the basic plan of campaign, with variations only in details. One problem which the Report had left unresolved was whether a private or a publicly-quoted company should be sold first. Another decision to be made was whether or not several companies should be sold simultaneously. The crucial factor was the behaviour of the institutions, and the Consortium had to decide when and how to approach them.

One of the first points the Sub-Committee had made in its Report was that the sale of the equity of the large, publicly-quoted companies 'must proceed concurrently with the resale of former privately-owned companies, e.g. English Steel and

Guest Keen Baldwins.'[1] The Issuing Houses insisted that they could not commit themselves to the first offer for sale until it was known that negotiations for the sale of the major privately-owned (i.e. non-quoted) companies were in an advanced stage, and all of Morison's pleas failed to move them.[2] This point, in fact, had initially been brought up by the Prudential and the Pearl Assurance, not by the Issuing Houses.

Nowhere were reasons given for this insistence of the institutions, but it is possible to speculate. First of all, there was the question of setting a price for the equity of the first company to be floated: if a price had already been agreed for a private sale, it would be easier to determine the right price for the first public offer. Secondly, long-term investment institutions are by the nature of their liabilities and responsibilities bureaucratic and conservative, and they dislike being first: the City is not unfamiliar with the saying that the institutional preference is to be the third buyer, even at a higher price, than the first. And thirdly, account must be taken of the deep and fundamental distrust which the institutions feel for the Issuing Houses. The characteristic of the Issuing Houses has always been to prosper by using their own brains and other people's money, and the institutions, with responsibility for the funds of their pensioners or shareholders or policy holders, react with great care to proposals of the Houses. Therefore, again, if a private sale price had already been negotiated, the institutions would have something against which to measure the price which would be set for the first public sale.[3]

The Governor, who had some sympathy with Morison's position, tried to get the Houses to modify their stance, but he found that some of their Directors, such as Kindersley of Lazards, felt very strongly about it as well. At a meeting of the Consortium on 8 June, the Governor warned them that insistence on the point might delay the first offer until October, and the Houses agreed to defer a final decision until the institutions had been

1 'Report by the Sub-Committee on Steel...', 13 May 1953, File 1953 Denationalisation, Box 369, MGP.

2 'I am sorry to note from page 2 of the Report that your Committee are still of the opinion that contracts for the sale of some of the larger formerly privately owned Companies must be signed before the first public offer. This will, I fear, delay the issue of the first public offer for quite some time. This in turn might have unfortunate repercussions on the whole operation.' Morison to Erskine, 16 May 1953, File 1953 Denationalisation, General, Box 369, MGP.

3 I am grateful to Jeremy Wormell for these suggestions.

consulted. The latter apparently held to their position, and by early July it was therefore clear that the first public offer for sale could not take place before the end of September or the beginning of October.[4]

Meanwhile, Morison had initiated negotiations over several companies, and by mid-September had fourteen in hand. On 13 July 1953 the new Iron and Steel Board would come into existence, and Morison wanted to announce at a press conference that negotiations were in hand for Guest Keen and English Steel to take back companies, and were far advanced for the flotation of Dorman, Long and Colvilles. The Consortium, however, thought that this would be 'most unwise'.[5] Morison pushed on, and by the end of July, as *The Economist* put it, 'The unscrambling of the steel industry has begun on a low flame with the private re-sale of Templeborough Rolling Mills to its former owners.' Templeborough had recently been very profitable, a factor which must have affected the negotiations, which saw some hard bargaining. Before nationalisation, Templeborough's 216,003 shares had been held one-third by British Ropes, Ltd, one-third by United Steel and one-third by William Cooke & Co., a joint subsidiary of British Ropes and United Steel. On 23 July 1953 the Treasury authorised the resale of 72,001 shares each to British Ropes and William Cooke at £7 5s. per share (the takeover price had been £7 per share); the one-third owned by United Steel had always remained with the company, rather than being vested in the Iron and Steel Corporation, and would be sold with United Steel when it was denationalised.[6] On 28

4 J.E. Beevor (of Slaughter and May), 'United Steel: Notes of Meeting of Consortium 24th September, 1953', File USC 1953 Docs/Memos, Box P377, MGP and Governor's Note, 10 June 1953, File G1/125, Bank of England Archives. Minutes of Meeting of Consortium, 8 June 1953, File 1953 Denationalisation, Box 369, MCP. Memorandum of Meeting held at Morgan Grenfell, 3 July 1953, Partner's File 29A, f.129, Barings Archives.

5 Memorandum of Meeting Held at Morgan Grenfell, 3 July 1953, Partner's File 29A, f.129, Barings Archives.

6 *The Economist*, 8 Aug. 1953, p.403. Treasury Minute, 23 July 1953, P.P. 1952–3, Cmd. 8922, Aug. 1953. Each side had a reason to bargain. The Agency would want a high price because of Templeborough's high profitability; on the other hand, it was being sold without its former holding in Scunthorpe Rolling Mill (which had gone to John Lysaghts). When announcing the sale Morison emphasised that the closeness of the takeover and resale prices was coincidental — that takeover prices were not relevant in fixing the prices at which steel

September 1953 the Treasury authorised another small, private sale, that of Round Oak Steel Works Limited to Tube Investments Limited.[7]

But what interested the City was not the sale of these small concerns, but that of the major unquoted companies, and this was considerably more difficult to arrange. On 24 September Morison joined a meeting of the Issuing Houses Consortium to thrash out the matter. He reported that on the previous day he had received proposals from Guest Keen Nettlefold and Vickers Cammell Laird which indicated that they were willing purchasers, but that there was a wide disagreement over price. 'He thought that ultimately he would succeed in doing a deal in each case but it would take two or three months. Consequently if the first public offer was to be postponed until completion of one of these two private negotiations there would be a long delay.' He was asked whether a statement could be published at the same time as the first public offer to the effect that these two firms were negotiating for the re-purchase of their former interests, but he pointed out that it would be rather embarrassing for him to ask a favour of the firms when he was negotiating with them — but that someone else could ask. Morison then left the meeting, and the representatives of the Houses discussed the matter. In the end, they decided that they would in principle be prepared to proceed with the first public offer on two conditions: one, that some satisfactory statement could be simultaneously published about the negotiations with Guest Keen and Vickers, and two, that the Prudential and the Pearl agreed.[8] Apparently they did, because final preparations for the first public sale went ahead.

The assumption had always been that the institutions were the key. Contacts had been postponed more than once, and by the time the Report was issued in May all that had been done was some careful soundings of the insurance sector, especially of the Prudential and the Pearl by the Governor, together with conversations with the head of the British Insurance Associa-

assets would be re-sold. This was a considerable change from his early position.

7 The Agency held all of the securities issued by Round Oak: £100,000 6% Cumulative Preference Stock and £635,265 Ordinary. They were sold for £1,603,574. Treasury Minute, 28 Sept. 1953, P.P. 1952–53, Cmd. 8962, Oct. 1953.

8 J.G. Beevor, 'Notes of Meeting of Consortium 24th September, 1953', File USC 1953 Docs/Memos, Box P377, MGP.

tion. The Consortium finally decided that after the announcement of the membership of the Holding and Realisation Agency, the time would be ripe for more widespread personal contacts. This would have two purposes, 'to bring them into the picture and, at the same time, obtain some idea of the terms on which it might be possible to do the first operation.'[9] A meeting was accordingly called of the Issuing Houses on 8 June 1953 to decide how best to approach them.

A list of the Managers and Secretaries of investment trusts had been drawn up and allocated to various members of the Consortium, with a few left to be handled by the consortium of brokers. Barings, for example, would approach Robert Fleming & Co. Ltd, Morgans would approach Philip Hill, Higginson & Co., and Bensons would approach a whole range of Edinburgh firms; in all there were fifty-three to be seen. In addition there were forty-six leading insurance companies which the Consortium would contact, and these were allocated to Houses according to whether they had a special relationship. For example, Morgans would see the Royal Exchange Assurance, of which Bicester had been Governor for fifty years. However, it was also decided that these contacts could not be made until the consortium of brokers had been formed and the Agency had been appointed.[10]

The announcement of the membership of the Agency was made two days later, with the only addition to the names as listed by the Governor and Morison on 28 April being that of C.P.L. Wishaw, a partner in Freshfields. The decision was then taken to put off the visits to the institutions until after 13 July 1953, when the Iron and Steel Board took over supervision of the industry from the Corporation and the denationalisation process would formally begin. Three members of the T.U.C.'s General Council had accepted posts on the Board: Andrew Naesmith, General Secretary of the Amalgamated Weavers Association and a Director of the Bank of England; J.O. O'Hagan, General Secretary of the Blastfurnacemen; and Evans, now Sir Lincoln Evans.[11]

Evans had accepted his knighthood from the Conservatives in the 1953 New Year Honours and thereby caused intense anger

9 Smith to Colville, 29 May 1953, File 1953 Denationalisation, General, Box 369, MGP.

10 Meeting of Consortium of Issuing Houses for Steel, 8 June 1953, File 1953 Denationalisation, Box 369, MGP.

11 Minutes of Meeting Held at Morgan Grenfell, 3 July 1953, Partner's File 29A, f.129, Barings Archives. *Financial Times*, 25 June 1953.

and revulsion on the left of the Labour Party. He was attacked in the 16 January 1953 issue of *Tribune* for accepting an honour at the moment when the main measure before Parliament was the denationalisation of steel. This attack exacerbated the already bad relations between the General Council and the National Executive Committee of the Labour Party; the General Council protested about the attack on Evans to the NEC, and the right-wing members of the General Council, which included Evans, were then virulently attacked by Aneurin Bevan, who had in August 1947 insisted on the nationalisation of the iron and steel industry.[12] The objections of the left to the appointment of Evans on 22 May 1953 as Vice-Chairman of the Board therefore came as no surprise. The General Council, however, strongly supported Evans and his two colleagues. On 24 June the General Council decided that not only were its three members entitled to accept posts on the Board, but were carrying out TUC policy by doing so. The classic text cited in support of this position was the statement made by Walter Citrine in 1938 when the Unemployment Assistance Board was set up: 'it is far better to exert our influence to control these things rather than to stand outside and to have to take the consequences.' Indeed, the press report went on to reveal that as far back as December 1951 the TUC Economic Committee (which Evans chaired) had known that trade unionists might be asked to serve on the Board, and at no time during the discussions did members of the General Council, affiliated unions, the Labour Party executive or the Parliamentary Labour Party object to the proposal. In fact, the General Council had sought to make it as easy as possible for trade unionists to serve.[13]

It had been decided early in the year that no operations could take place until the membership of the Board was known, and particularly the trade union membership. From the City's point of view this could now hardly be bettered, and after 13 July the Issuing Houses could, with more confidence, approach the institutions. It was Morgan Grenfell's duty to visit the most important of the insurance companies, the Prudential, and once this had been done, Smith and Erskine reported to the Governor and Mynors. 'The Pru [were] against both a powerful syndicate

12 Michael Foot, *Aneurin Bevan 1945–1960* (London: Davis-Poynter, 1973), pp. 392–93.

13 *Financial Times*, 25 June 1953. Morgan, ed., *Backbench Diary*, entry for 26 May 1953, pp.234–35. Naesmith was also knighted in 1953.

to take over the bulk of issues, and any attempt to do the major companies simultaneously.' They believed that the insurance companies should not take more than 50%, with the Prudential taking one-fifth of the sector's total. They might, the Prudential had added, have reservations about buying shares in some of the weaker companies. The Governor accepted the concern of the insurers that they should not have a dominant shareholding in the steel industry, but he hoped that they would keep their position open so that if the first issues were a success, some of their unused resources could be called on for later issues. Smith pointed out that '50% was somewhat below the insurance companies' proportion of existing underwriting lists.'[14] He suggested that it might be possible 'to get some help from large industrial companies interested in the maintenance of free enterprise', but no names were mentioned. The Governor told them, for their private and personal information, 'that the Bank might be prepared to find £1 or £2 million, if this were necessary and helpful in completing an underwriting list from time to time.' They then agreed that the Governor should not himself see insurance company chairmen: rather, the members of the Consortium would set out on their rounds, beginning with those institutions where they had representatives on the Board.[15]

When the Issuing Houses Committee met on 24 August, they reported a general impression that support from the insurance companies would cover no more than 50% of the first flotations,[16] although they would judge each on its merit — the position they had always taken. 'It was clear that the more difficult task would be to find guarantors for the remaining 50% of the early Issues.'[17] By the time of the first public issue, however, sufficient institu-

14 This, presumably, was the result of Labour's threats to nationalise the insurance companies: they wanted to keep their heads down and not be open to charges of having 'Prudentialised' the industry.

15 H.C.B.M., 'Steel', 16 July 1953, File G1/125, Bank of England Archives.

16 Following a meeting with the Pearl, the Investment Manager telephoned Erskine 'that they thought the sort of figure they would be willing to take was 2$^1/_2$% of the total of each issue or, expressed in another way, 5% of the half which they thought should be the maximum for the insurance companies as a whole. This compares with the figure of 20% of the half recently mentioned by the Prudential when we had our meeting with them.' R.G.E., 'Steel', 27 Aug. 1953, File 1953 Denationalisation, Box 369, MGP.

17 Minutes of Meeting Held on 24 Aug. 1953, File 1953 Denationalisation, Box 369, MGP.

tional support had been promised: what was more problematical was whether they would then hold them or unload them.[18]

The third problem which the Issuing Houses Consortium had to resolve was whether or not a group of companies should be brought to market simultaneously. If a single company was to test the market, Morison would decide which it was to be. Morison had based his early planning on the assumption that six to eight of the companies would be sold during the summer of 1953, but by late February of that year Forbes for one thought it unlikely that more than one or two individual operations could take place before the summer holidays. Thus ideas were fluid, and various permutations were put forward, discussed and abandoned.

Meanwhile, following Morison's suggestion of February 1953, those Houses which had the major publicly-quoted companies as clients worked to prepare the documentation so that whichever company was selected would be ready. This task was not evenly spread: as Mynors noted in a talk with Morison in late April, Morgan Grenfell 'now have on their list six companies, including South Durham, United Steel, Colvilles, Dorman Long and Stewarts & Lloyds (jointly). This is too long a list for any one House and will need very careful dressing up.'[19] The Governor, too, thought this might be tricky, as indeed did Bicester, who mentioned to the Governor that 'Morgans are getting an almost embarrassing lot of this business.'[20] The recompense would be prestige rather than fees, which would in any case be evenly divided amongst the Consortium. On 18 June, when permission was given for the Issuing Houses Committee to study the schemes for Stewarts & Lloyds, United Steel, John Summers and South Durham, all four had presumably been prepared by Morgan Grenfell.[21]

18 The Sun Insurance wrote on 13 Oct. 1953 that 'It was our unanimous decision to accept the underwriting offer of 250,000 shares. Bearing in mind, however, the Board's wish that £500,000 of Steel Ordinary Stock should be our *ultimate* maximum, we considered how much of the United Steel we would like to hold permanently. We thought that something between 75,000 and 100,000 shares would be appropriate. We visualise, however, that we may be left with considerably more than this and will therefore be sellers." Eric ? to Francis ?, 13 Oct. 1953, File USC 1953, Box P377, MGP.

19 H.C.B.M. 'Steel', 22 Apr. 1953, File G1/125, Bank of England Archives.

20 Governor's Note, 1 May 1953, File G1/125, Bank of England Archives.

21 K.C. B[arrington], 'Sub-Committee on Steel. Possible items for the

A few days later, two of the members of the brokers' consortium, Cazenove Akroyds & Greenwood and Hoare & Co., submitted a memorandum arguing that there were 'difficulties inherent' in the Report of the Sub-Committee. They believed that 'a succession of Steel Equities separately guaranteed is only possible in an atmosphere of mounting success. It may be very difficult to overcome the attitude of "wait and see".' Therefore, they had come up with a scheme to 'ensure at the outset that arrangements have been made to cover the whole, or at any rate the majority, of the operation.' They suggested two possible methods. In the 'Package' method, the equity of the seven or eight leading companies, amounting in cash terms to about £70 million, would be underwritten as a composite unit.[22] The second scheme was based on the principle of a revolving credit, which would be available during the whole of the operation; a Guarantor's Syndicate would be formed, which would last for at least two years.[23]

Morgans drew up a list of arguments against the suggestions. The combined equity of the first seven companies might be as much as £100 million: the Consortium anticipated difficulties in the market's finding that sum over a period of months, let alone for a single operation. If the companies were sold simultaneously, the insurance companies would probably have to guarantee more than 50% against an unknown response from the public, and they had steadfastly refused to contemplate taking more than the bare 50%. A simultaneous issue would probably have to be made more attractive, i.e., with the share price lower, than a single one, thereby lessening the chances of the Agency's making the profit on which it was insisting. It would be difficult to know how to determine the yield for each company. Finally, to get the first seven issues organised would inevitably

Agenda of future Meetings', 18 June 1953, File 1953 Denationalisation, Box 369, MGP.

22 The idea was that underwriting would be arranged in units of, say, £1,000 each. Each unit might consist of, say, 300 Stewarts & Lloyds, 100 Dorman Long, 50 Whitehead and so forth. Each company would then simultaneously publish a prospectus and a market would develop in each company's shares. 'More Detailed Notes on the Package Scheme', n.d., and 'Memorandum from Cazenove Akroyds & Greenwood & Co., and Hoare & Co. on Steel De-nationalisation', 29 June 1953, for the quotations, both File 1953 Denationalisation, Box 369, MGP.

23 *Ibid*.

mean delays, and their sale might not then take place until after Christmas. In short, a single first issue would be a good test.[24] A week later the Prudential came out firmly against both suggestions,[25] and the idea was dropped. Thereafter, the only question was which company would be the first.

For some time it had been widely forecast that it would be Stewarts & Lloyds, possibly because it was so profitable: in 1952/3 its gross profits were over twice those of United Steel (£13.4 million against £6.1 million), while its net income in 1952 was over two-and-a-half times that of United Steel (£5.57 million against £1.96 million).[26] Indeed, the Report of the Issuing Houses Sub-Committee in May had so recommended. On 30 June 1953, however, the *Financial Times* reported that while negotiations with half a dozen companies had reached an advanced stage, it was now regarded as unlikely that Stewarts & Lloyds remained at the top of the list. No reason was given. On 9 July Morison wrote to Erskine that he hoped that Morgans could settle the capital structures of United Steel and Summers before the holidays, so that the necessary mechanical work could be ready by early September, the first evidence that he had nearly decided that United Steel, the largest of the companies, should undergo the first test.[27] At a meeting of the Issuing Houses Consortium on 24 August 1953, Erskine reported that Morison thought that United Steel would probably 'form the subject of the first Offer.' He so decided, and on 9 October submitted to the Treasury the formal proposal to sell; the Treasury, which had earlier agreed to the proposed capital reorganisation of United Steel, gave its formal consent on 12 October 1953.[28] Only the final, formal preparations remained to be made.

24 'Arguments Against Doing the First Seven Companies Simultaneously', 6 July 1953, File 1953 Denationalisation, Box 369, MGP.

25 H.C.B.M., 'Steel', 16 July 1953, File G1/125, Bank of England Archives.

26 *The Economist*, 20 June 1953, p.830.

27 File 1953 Denationalisation. General, Box 369, MGP. The United Steel Companies Limited was made up of Appleby-Frodingham Steel Co., Scunthorpe; Distington Engineering Co. Ltd., Workington; Samuel Fox & Co. Ltd., Sheffield; Steel Peach & Tozer, Sheffield; United Coke & Chemicals Co. Ltd, Treeton; United Steel Structural Co. Ltd., Scunthorpe; United Strip & Bar Mills, Sheffield; Workington Iron & Steel Co., Workington; and Yorkshire Engine Co. Ltd., Sheffield.

28 Minutes of a Meeting Held on 24 Aug. 1953, File 1953 Denationalisation, Box 369, MGP. Treasury Minute, 12 Oct. 1953, P.P. 1952–3, Cmd. 8977, Oct. 1953.

Bringing United Steel to Market

The period from February through October 1953 saw the answers to two separate but interrelated problems being worked out: firstly, how to sell the industry, and secondly, how specifically to bring the United Steel Companies to market. The work of the Issuing Houses Sub-Committee on Steel was intended to establish the general principles for all of the sales, in addition to which the Bank of England and the Realisation Agency had to settle questions about pink forms, gilt-edged stock and other details. At the same time the specific problems of re-organising the capital structure of United Steel and drawing up its prospectus and attendant documents were being tackled. This was a matter of some importance, since it was vital for the whole programme that the first flotation be a success. Indeed, the decisions on details of the United Steel flotation sometimes established the general principles to be followed in successive sales.

Near the end of January 1953 Morison told Hunter that he hoped to begin talks with companies at an early date, and Hunter alerted them. United Steel had already prepared one memorandum, and would draw up another, on the finance which the company required for past and future development. When in late February, Morison told Walter Benton Jones, the chairman of United Steel, that his company and Morgans ought to start planning for its sale, R. Peddie, the Secretary of United Steel, sent the memoranda to Erskine. In early April Morgans sent a memorandum to United Steel setting out the minimum conditions upon which negotiations with the Realisation Agency should begin. By the beginning of May Morgans and United Steel presented Morison with a plan suggesting how they might re-organise the company's capital structure to provide the basis for flotation. Negotiations went on for some weeks, until in mid-July, Morison made his formal counter-proposals. These were finally accepted by the United Steel Board on 16 September 1953, enabling Erskine to report to the Issuing Houses on 24 September that United Steel would be the first company to be sold. The terms were announced to the press on 15 October, and on 25 and 26 October the Prospectus was published in a large number of national and regional newspapers. The lists opened on 29 October and closed on 9 November.

When United Steel was nationalised, the procedure had been for the Iron and Steel Corporation to 'buy' United Steel's Ordinary and Preference shares, paying for them with £21.3 million of $3^{1}/_{2}$% Steel Stock.[29] The Iron and Steel Corporation now wanted to sell the company back to the investor, and selling at the 'takeover' price would mean selling it for at least £21 million. In theory it would be possible to sell it with the same capital structure[30] as it had had in 1949, a ratio of ordinary shares to prior charges of two to one (£10.2 million in equity and £5.8 million in preference shares). However, this was never seriously contemplated, partly because the company's assets had grown since the pre-nationalisation share structure had been established and partly because a much higher return than that suggested by the takeover price had to be made from the sale. Therefore, once Benton Jones had been alerted that events were in motion, the company drew up two memoranda detailing the funds the company would require and proposing a possible capital structure.

29 Compensation price was that quoted on the Stock Exchange in Oct. 1948:

Ordinary Stock	10,194,000 shares @ 30/4 =	£15,460,900
Preference	5,097,000 shares @ 22/10 =	£5,819,075
	Total compensation	£21,279,975

The assets per share of the 1948 balance sheet date was 38/-. The adjusted 1953 value, assuming payment of a dividend of 8%, was 45/-. W.B.J., 'Provision of New Money Required to Cover the Cost of the Extensive Developments Undertaken by the Company Since 1945', 10 Jan. 1953, File USC 1953, Docs/Memos, Box P377, MGP.

30 The capital is represented by ordinary shares and the various types of prior charge. Ordinary shareholders own the participating capital: they can vote, and own the profits and assets after the rights of the prior charges and other creditors have been met. Prior charges are made up of preference shares, debentures and loan stocks. These have a 'prior' right to the income and assets if the company gets into difficulties. The exact position of Preference shareholders are regulated by the individual trust deeds. In general, however, they have a maximum rate of dividend, which is regarded as fixed, but may be cut if there are no profits. In the case of a cumulative preference share, unpaid dividend rights accumulate until the company can pay, whereas in the case of the ordinary shareholder, a reduced dividend is lost forever. If the preference share is redeemable, it can or must be retired by the company on or between given dates, with the exact conditions again depending on the trust deeds. Debentures should be regarded as representing a loan to the company and are usually secured against specific items amongst a company's assets; a loan stock is also a loan, but is usually only secured against the assets in general. Again, rights of the loan stockholders and their position vis-à-vis bank debt and other creditors depends on the specific terms of each trust deed.

In the first memorandum the company concentrated on establishing how much money it needed and why. For capital expenditure under way in September 1952, plus developments planned for the period to September 1957, the company estimated that it required £33 million. Of this £17 million could be generated internally, leaving £16 million to be found externally; £10 million would be required in 1953, £5 million in 1954 and £1 million in 1955. The company admitted that £16 million over two-and-a-half years was large in relation to issued ordinary share capital (£21.3 million) and suggested that instead, it be considered in relation to fixed assets. In 1948 the ratio of ordinary share capital to fixed assets (including investments) had been 5:7, whereas in 1952 it was 5:11, leaving a good margin to increase the number of ordinary shares. On the other hand, loan capital had become a cheaper source of finance than ordinary share capital, so that it was advantageous to raise as much money as possible in the form of loans. Yet the Agency might consider a large proportion of loan capital unsuitable, and therefore might consider it preferable to capitalise reserves,[31] in that shares might be easier to sell (i.e., each share would cost less) and the amount of loan

31 A company can have authorised share capital and issued share capital. The latter is already issued and in the hands of shareholders; the former has been authorised by the shareholders but not yet issued. In 'capitalising the reserves', the company issues new shares *pro rata* to existing shareholders to reflect the increased value of its assets. This would normally leave the shareholder unaffected, since the price of existing shares would fall (reflecting lower assets and earnings per share), but the number of shares held would rise to compensate the shareholder exactly. In this case they would be sold to the Agency, which would then sell them on the market. As Erskine wrote on 8 October 1953 to S.S. Wilson, the Secretary of the Holding and Realisation Agency, 'Our experience is that it is most desirable that a large public company should have a reserve of unissued shares available for issue at the discretion of the Board. Such shares may be required from time to time either in connection with the acquisition of another undertaking or for issue for cash to raise funds for capital expenditure. If the Board has no reserve of unissued shares at its disposal, an interval of several weeks must ensue while the necessary extraordinary general meeting is convened and the resultant delay and publicity may have a prejudicial effect upon the negotiations or upon any arrangements for marketing shares. It is therefore our practice to recommend companies, for whom we act as financial advisers, to take a convenient opportunity to increase their authorised capital as necessary to ensure that a reserve of shares is available.' File USC 1953, Box P377, MGP.

might not appear so disproportionate.[32]

In the second memorandum the company considered how the money might be provided. First of all, the increase in assets between 1937 and 1952 — £64,534,000 — would justify doubling the share capital. The existing ratio between ordinary and preference capital was 2:1. The existing preference issue might be converted into ordinary shares, but since the dividend was only 4 1/2%, it would be better to keep it and have another issue (which would have to carry a dividend of about 5 1/2%). An alternative might be to increase the ordinary capital to £20 million without increasing the number of preference shares, with the balance being provided by an issue of five- or ten-year unsecured loan stock. If at all possible, 'debentures should be avoided.'[33]

A fortnight after the second memorandum was written, Benton Jones met with Morison, who said that United Steel and Morgan Grenfell should prepare a resale scheme as quickly as possible. Peddie immediately sent the memoranda to Erskine, and a week later Erskine, Benton Jones and J. Ivan Spens met for preliminary talks. For the following three weeks Erskine and Kenneth Barrington studied the previous five years' financial accounts, took account of the company's financial forecasts, considered what might be the best way to meet the company's needs, and then drafted a memorandum setting out a scheme to serve as a basis for opening negotiations with the Agency.

First of all, £2 million had been borrowed from the Iron and Steel Corporation during the financial year ending in September 1952; the company estimated that another £6 million would need to be borrowed from the Corporation to finance capital expenditure up to 30 September 1953, about which time the sale

32 W.B.J., 'Provision of New Money Required...', 10 Jan. 1953, File USC 1953, Docs/Memos, Box P377, MGP.

33 W.B.J., 'Suggestions for the Provision of New Money', 13 Feb. 1953, File USC 1953, Docs/Memos, Box P377, MGP. The stricture against debentures apparently required no explanation, but there were at least three reasons why a borrower always tried to avoid them. First of all, they were secured against specific assets, such as certain plant or land, thereby constraining the freedom of the company in dealing with those assets. Secondly, debentures often had tighter ratios than loan stock. These ratios, set in the trust deed, controlled the amount the company might additionally borrow. And thirdly, if the company was unable to meet the interest payments, the debenture holders, unlike the preference shareholders, could call in the liquidators and close the company down. Elucidation by Jeremy Wormell.

of United Steel's equity might take place. Therefore, the Memorandum would assume that the Corporation would have made loans of £8 million to United Steel, which would have to be repaid. Further, the company estimated that they would require another £7 million in 1954; it was to be assumed that individual steel companies would be denied access to the capital market while the Agency was trying to sell the rest of the industry, and therefore the company should arrange for the provision of the £7 million before its equity was sold. One way of meeting part of this requirement was for the company to sell its 40% stake in the steel company John Summers & Sons Ltd to the Agency.[34] They held £2,805,822 Ordinary shares; the shares in the hands of the public had been taken over by the Corporation at 34/- [£1.70] per £1 unit; and on the assumption that the Agency would try to recover the book value of its holding in John Summers, the shares held by United Steel were worth about £5 million. The alternative to selling this holding was a very large amount of fixed charges, since the Agency needed to limit the amount of United Steel equity it had to sell, for fear of overloading the market. If the Summers stock was not sold while United Steel was still nationalised, there would not be another opportunity for two to three years, since the company would hardly be allowed to sell a large block of privately-owned steel shares while the Agency was trying to unload all of the publicly-owned ones. Morgans assumed that United Steel would follow this advice, with the result that the requirements of the company then dropped from £36.3 million (Morgans ignored the £1 million required in 1955) to £31.3 million. On the assumption that the Agency would seek to recover at least the book cost of its investment in United Steel, the scheme had to provide for a market capitalisation of about £32 million.

First of all, there was the £5 million 4 1/2% cumulative prefer-

34 United Steel's stake in John Summers dated from Nov. 1938, when John Summers needed £3.6m to complete a strip mill. Unfortunately, the concurrent difficulties of another steel firm, Richard Thomas & Co., made it impossible to raise such a sum from the public. In the circumstances, the Bank of England provided £1m (through the Bankers' Industrial Development Co.), United Steel (at the Bank's request) put in £1.2m and took a 1/6 stake in John Summers, and John Summers raised £2m through a debenture issue sponsored by Morgan Grenfell and Helbert, Wagg. Sayers, *Bank of England*, pp.549–50. File John Summers & Sons Ltd Issue of £2,000,000 4 1/2% First Mortgage Debenture Stock 1939, MGP. Burk, *Morgan Grenfell*, Chapter 6.

ence stock. Because of its low dividend, it could not be sold at par, and Morgans assumed it was worth 16/- [80p] per share, or £4 million. The capital to be raised through the sale of ordinary shares would then be adjusted *pro rata*, and this would determine the amount of new 6% preference shares which would have to be issued. The company should have the right to redeem this new preference issue; not only was a 6% dividend high, but it was liable to a distributed profits tax of 20%.[35] Morgans therefore proposed that to provide new finance, the Agency, before selling the equity, should subscribe for £1.5 million 6% redeemable preference shares at par and for £8.5 million 5% debenture stock 1968/78 at par, to be secured by a floating charge.[36] It was suggested that the Agency should pay up the preference shares in full and make a part-payment on the debenture stock, with the balance available to be called up as and when the company required the capital.[37]

35 Labour Governments had a bias against the distribution of profits, believing that they should be ploughed back into the company. Morgans and United Steel could only hope that the Conservatives would repeal the tax.

36 A fixed interest security secured by a floating charge was an innovation recently pioneered by Erskine and Morgan Grenfell. Basically, instead of the debenture being secured on a specific asset, it was secured on the total assets of the company, and thus left the company freer to manage its affairs.

37 In their May Report, the Sub-Committee had cited as one of Morison's intentions that 'It would be preferable to have prior charges in the form of redeemable preference shares in order to keep loan capital in reserve for new money. If new money is required in the immediate future, that may have to be provided by the Agency in the form of loan capital before the sale of the equity.' When the first drafts of the Offer for Sale were discussed, Morison was apparently unhappy at what appeared to be an unpredictable commitment to subscribe loan stock 'as and when required by the Company to do so'. Draft of 10 June 1953 of Suggested Specimen Skeleton Offer for Sale, File Denationalisation, General, Draft Docs, Box 369, MGP. 'Morison . . . recognises that it will be necessary to ensure that each Company will be sufficiently supplied with its cash requirements for the next two or three years so that it should not find itself having to raise money on the market while the Agency is completing its work. This may involve the subscription by the Agency of debenture stock or possibly, in appropriate cases, short-term notes (say five years) or long-term notes (say fifteen to twenty years). In each case, however, I think that the debenture stock or notes will be issued and subscribed even if only partly paid in the first instance.' Wishaw to Beevor, 17

United Steel had suggested that the issued capital be brought more into line with the assets employed in the business. To do so, Morgans proposed that the sum of £863,454 in the Share Premium Account,[38] together with £3,845,546 from other reserves, should be capitalised and 4,709,000 new ordinary shares credited as fully paid issued to the Agency. If the sale of United Steel's stake in John Summers was carried through, there would be a realised profit of nearly £3 million, which could be turned into reserves and then used to increase the number of shares in issue by a further capitalisation, leaving the balance to be found from General Reserve. Finally, to round off the ordinary and preference capitals, £97,000 of the existing 4$^1/_2$% preference shares should be converted into ordinary shares. The capital structure then became as follows:

5% Debenture Stock 1968/78	£8,500,000
4$^1/_2$% Cumulative Preference Shares	5,000,000
6% Cumulative Redeemable "A" Preference Shares	2,500,000
Ordinary Shares	15,000,000
	£31,000,000[39]

The question now became, how much would the Agency earn from the proceeds of the sale? As far as the ordinary shares were concerned, much of the answer would turn on the yield basis;[40] Morgans assumed that the ordinary shares could not be sold on a yield basis of less than 7%, and because the existing dividend was 8%, the stock could not be sold much above par. On the other hand, 'from the point of view of labour relations it may be undesirable to raise the dividend rate above 8%.' A dividend of 8% on ordinary shares sold at 23/-[£1.15] would give a yield of 6.96%. The proceeds of sale would then be as follows:[41]

June 1953, File Denationalisation, General, Box 369, MGP.

38 For the purpose of the Companies Act the Share Premium Account was part of United Steel's permanent capital.

39 Morgans also suggested that the company pay the Agency a special dividend of £1m tax-free from undistributed profits, which would be immediately used to subscribe at par for a further 1m 6% redeemable "A" preference shares. The procedure was necessary to avoid any liability at a later date for distributed profits tax if the preference shares were subsequently redeemed.

40 The yield basis of an equity is the dividend in pence payable per share divided by the price of a share.

41 *Capital Cover*:

Net assets per Consolidated Balance Sheet 27 Sept. 1952	£29,716,164
Add: New finance 1953	£10,000,000

£15,000,000 Ordinary Shares at 22/- (after deducting 1/- for expenses)	£16,500,000
2,500,000 6% Redeemable "A" Preference Shares at say 21/-	2,625,000
£5,000,000 4½% Cumulative Preference Shares at say 16/-	4,000,000
£8,500,000 5% Redeemable Debenture Stock 1968/78 at par	8,500,000
	£31,625,000

This, as Morgans noted, would provide the Agency with only a very small surplus of £345,025 towards losses on the less attractive companies and to reflect retained profits since nationalisation.

Morgans then explained the basis of some of their recommendations. First of all, at a time of such high interest rates, institutional investors (who took up most prior charges issued by first-class companies) tried to lend for as long a period as possible and to restrict the right of the borrower to convert the securities to a lower rate of interest at an early date. The borrower, conversely, would want to have the right to convert if lower interest rates returned. In this situation the fact that the Agency would hold the prior charges should be advantageous, since Morgans assumed that it would be open to the company to redeem any of the prior charges for cash at par before they were sold to the public. The company would like to have a larger equity capital than was possible at the flotation, but after denationalisation it might be possible for the company to make a rights issue and use the proceeds to redeem some of the prior charges held by the Agency.

Morgans pointed out that it was difficult to sell preference

Retained profits to Sept. 1953 (say)	£2,000,000
Capital profit on sale of Summers (say)	£3,000,000
	£44,716,164
Deduct: Debenture Stock (which is covered 5¼ times)	£8,500,000
	£36,216,164
Deduct: Preference Capital (Preference Capital + prior charges covered 2.8 times)	£7,500,000
Net assets available for Ordinary Shares	£28,716,164
"Break-up" value of Ordinary Shares	38/4 per £ unit

shares, and if the 6% preference were issued, it would be necessary to provide that they could not be redeemed by the company before, say, 1964, and even then only at a stiff premium. However, the option at that date meant that they would only be redeemed if it were advantageous to the company. But expensive as preference capital was, it reduced the amount of loan capital required. The company would have to raise a large sum either through debentures or unsecured loan stock; however, this would minimise the burden of distributed profits tax as well as limiting liability for excess profits levy. Morgans also pointed out that it made little material difference to the company's ability to conduct business whether they borrowed on a debenture stock secured with a floating charge or on unsecured loan stock. Further, a 5% debenture stock was the cheapest possible form of borrowing.

In concluding, Morgans reiterated that their scheme was the minimum likely to serve as a basis for negotiations with the Agency, and undoubtedly the Agency would press for better terms. However, the amount which the company would have to allocate to service the share and loan capital (including debenture interest and distributed profits tax on preference and ordinary dividends) represented only 38% of the average available net profits from 1948 through 1952. Consequently, there was scope for increasing the amount of the preference capital, if the Agency insisted, without reducing the retained profits below 50% of the net profits, which was probably the minimum which should be retained.[42]

On 27 April 1953, Benton Jones and Peddie called on Morgans to tell them that, in general, the United Steel Board accepted the Memorandum as the basis upon which the Agency should be approached, although 'for reasons of sentiment' some members of the Board hated to part with the stake in Summers. The only change the Board suggested was £10 million for the debentures, rather than £8.5 million, on the grounds that the former figure would give a better cash margin in the early years. Peddie undertook to redraft the Memorandum in a form suitable for submission to the Agency, and after it had been agreed with Morgans (and a Chairman's letter had been drafted and the Auditors had produced a draft of their Report), a formal approach

42 K.C. Barrington, 'The United Steel Companies Limited', 2 April 1953, File USC 1953, Docs/Memos, Box P377, MGP.

would be made to Morison.[43]

The re-drafted Memorandum differed little from Morgan Grenfell's except that it was much less detailed; the only change was that in their figures United Steel included the £1 million required in 1955. The Memorandum was then sent to Morison, and discussed at a meeting on 5 May 1953 between Morison, Benton Jones, Peddie and Erskine.[44] One drawback of this scheme for the Agency was that it did not provide a 'loading' — i.e., a surplus over the actual requirements which could compensate for the losses which were expected from selling the weaker companies, or produce a profit for the Government. The United Steel Board suggested several ways of providing a higher price for the Agency, with their preference being a scheme which left the capital structure unchanged but increased the dividend on the ordinary shares to 10%, thus allowing the price to be increased and raising an extra £4 million. Morison was still not satisfied and submitted his own proposals to the company, and in return the company prepared other schemes. These they sent to Erskine, writing that they would 'be grateful if you would put forward your own views on the form of capital structure which would best suit the Company's interest and meet, in so far as it is necessary to do so, Sir John Morison's demand for a substantial profit on the transaction.'[45]

On 13 July 1953 Erskine and Barrington of Morgans met with Benton Jones, Peddie and Gerard Steel, Managing Director of United Steel, to discuss the resumption of negotiations with the Agency. Morgans believed that further negotiations should be based upon four considerations: (1) the company was not willing to forecast profits of more than £6 million[46] and debenture

43 K.C.B., 'The United Steel Companies Limited', 29 April 1953, File USC 1953, Docs/Memos, Box P377, MGP.

44 R.P., 'Proposals for Reorganised Capital Structures and Sale of the Company's Securities By the Iron & Steel Holding and Realisation Agency', 1 May 1953, File USC 1953, Docs/Memos, Box P377, MGP.

45 Memo by Barrington, 22 May 1953, File USC 1953, Docs/Memos and Peddie to Erskine, 11 July 1953, File USC 1953 for the quotation, both Box P377, MGP.

46 Benton Jones wrote to R.H.V. Smith on 3 June 1953 that profits to 30 Sept. 1953 might be £9m, rather than the £8m of an earlier estimate. File USC 1953, Box P377, MGP. It was emphasised that this level of profit was believed to be unusually high and would not be sustained in the following years. By Oct. 1953, demand for steel seemed to be faltering and export price premiums were running off. However, in

interest needed to be deducted in any calculations; (2) ordinary capital should represent approximately 50% of the total market capitalisation; (3) the amounts distributed should not exceed 50% of the profits after tax; and (4) the rate of dividend on the ordinary stock should not exceed 10%. The five discussed a possible scheme, but Erskine thought it would be a tactical mistake to go back to the Agency at this stage with a plan embodying a surplus to them of nearly £3 million: he urged a revision which would reduce this surplus by nearly £1.5 million by reducing the ordinary and preference shares in equal proportions. It was pointed out that if the Agency accepted their scheme in principle, there was a sufficient margin to adjust the surplus for the Agency, if necessary, by a moderate increase in the rate of dividend and/or an adjustment in the valuation of the prior charges, including if necessary an increase in the dividend on the existing preference stock to 5%. After more discussion, the amended scheme was accepted by Benton Jones as the basis on which they would resume negotiations. He and Erskine would see Morison on 24 July and submit the revised proposals to him. All the calculations depended upon the basis on which the ordinary shares and the prior charges were sold, and they and Morison knew that this depended vitally on the state of the market at the time. In any case, they would emphasise to Morison that the company would not wish to go beyond the revised proposals, which involved distributing roughly one-half of the estimated maintainable earnings before tax.[47]

Negotiations were resumed and continued for some further weeks. Finally, on 11 September 1953 Morison handed Erskine the Agency's proposal for the re-organisation and disposal of United Steel's capital:

£14,000,000 Ordinary Shares at 25/6 &£17,850,000	
4,000,000 $5\frac{3}{4}$% Redeemable Preference Shares at 20/-	4,000,00
5,097,000 $4\frac{1}{2}$% Cumulative Preference Shares at 16/-	4,077,00
10,000,000 $4\frac{3}{4}$% Debenture Stock at 98	9,800,00
	£35,727,00

£21.5 million was needed by the Agency to cover the takeover price and the proposal provided £11.3 million in new money for the company, leaving £2,927,000. From this it was necessary

Dec. 1954, United Steel announced a trading profit of £9,136,496 for the year to 30 Sept. 1954. *Financial Times*, 16 Dec. 1954.

47 Memo of meeting held on 13 July 1953, File USC 1953, Docs/Memos, Box P377, MGP.

to deduct the estimated cost of selling the ordinary shares, i.e., 1/- per share, or £700,000; in addition, if the preference shares were sold at a 1/- discount, another £455,000 had to be subtracted. In the end, the surplus accruing to the Agency would be £1,772,000.[48]

The proposal was considered by the Board of United Steel on 16 September 1953. Comparisons were presumably made between its terms and the requirements of the company as set out by Erskine and Peddie. Both had assumed that equity shares and prior charges had each to constitute 50% of the capitalisation: with the Agency's scheme, the £17,877,000 in prior charges came to 50.04% and the equity to 49.96%. The company had regarded a dividend of 10% as the maximum they were prepared to pay, and Morison had suggested 9%. With regard to the debentures and redeemable preference stock, the company had expressed a wish that the rate of interest on the former should be slightly below 5% — it became 4$^{3}/_{4}$% — and that on the latter slightly below 6% — which became 5$^{3}/_{4}$%. In short, the Agency had clearly tried to meet all of the company's main requirements, and accordingly the Board accepted the proposal.[49] The way was now clear for United Steel's equity to be floated; the Agency would hold the prior charges until 1961, when they too would be sold.[50]

While Morgan Grenfell and United Steel were concentrating on the capital reconstruction of the company, members of both firms, as well as of the Bank and the Consortium, had also to decide upon a number of other points. The intention had always been to establish a common pattern for all of the equity sales, and thus considerable amounts of time were spent in determining, for example, who should be eligible for pink forms, or just what gilt-edged stock could be exchanged for steel equities, or which newspapers ought to publish a full prospectus and which only a shortened version.

In their Report, the Issuing Houses Sub-Committee had stated that pink forms were to be available for the former holders of both the ordinary and preference share and loan capital of the companies, but the Bank's assumption in late May 1953 was that only ordinary shareholders would receive them. However, by

48 Memorandum handed by Morison to Erskine on 11 Sept. 1953, File USC 1953, Docs/Memos, Box P377, MGP.
49 Peddie to Erskine, 11 July 1953, File USC 1953, Box P377, MGP.
50 In effect, United Steel sold the prior charges to the Agency, receiving thereby the funds necessary for development.

10 June the decision had been made that pink forms would go to all former owners of share capital (including preference), but not to former holders of loan capital (whether debentures or loan stock.)[51] But there was a further consideration, and this was whether pink forms ought to go to steel company employees as well. United Steel thought not:

> With regard to the offer of shares to employees, we think that on the whole it would be better not to do this. When we were nationalised we took the line with our employees that this made no difference to the organisation of the Company and was really no concern of management or men in their internal relationships. We feel that the mere offer of pink forms would, in fact, be an invitation to subscribe and might be interpreted as a change in attitude. This would be undesirable.[52]

In any case, most employees would not lose out, as Smith explained to H.C. Drayton, chairman of a dozen investment trusts:

> The present intention is to state on the front page of the Offer for Sale that preferential consideration on allotment will be given to all applications for small numbers of shares which I would think, in the ordinary case, would result in employees being allotted more or less the number of shares they wanted.[53]

The pink forms themselves would have to come in two types, as indeed would non-pink forms, since applicants could pay either in cash or in gilt-edged stock.[54] This raised the question of which gilt-edged stocks would be eligible. The Bank and the Treasury had long discussions over this: at first it did not seem practicable to make the list very comprehensive, but gradually attitudes changed and it was decided to make the list as wide as possible. Certain quoted stocks were excluded, such as Victory Bonds and guaranteed stocks such as British Electricity, Gas and Transport, as well as unquoted Government debt such as Defence Bonds and Saving Certificates. In the end a list of twenty-eight gilt-edged stocks were eligible, ranging from Consols to

51 H.C.B.M., 'Memorandum', 26 May 1953 and K.F.C.[hadwick], 11 June 1953, both File 1953 Denationalisation, Box 369 MGP.

52 Peddie to Erskine, 22 Sept. 1953, File USC 1953, Box P377, MGP.

53 Smith to Drayton, 17 Sept. 1953, File 1953 Denationalisation, General, Box 369, MGP.

54 In the end, application forms came in four colours, azure and pink for cash, green and buff for stock.

War Loan to Steel Stock. Applicants could tender more than one issue, but a separate application would have to be made for each type. The price at which each issue would be accepted in payment would be announced by the Bank on 23 October 1953, and that price would be held for the whole time the lists were open,[55] thereby essentially freezing the gilt-edged market for the duration of the sale.

Possible applicants had to be reached, and the only way to do so was by extensive newspaper advertising. Full-page or abridged advertisements would be run in thirty-eight newspapers and four periodicals, the latter being *The Economist*, the *Statist*, the *Stock Exchange Gazette* and the *Investors Chronicle*. The *News of the World* and the *Sunday Pictorial* both refused advertisements because they were full. The Nottingham *Guardian Journal* and the Dundee *Courier* both complained bitterly at being left out, but Charles Barker and Sons, who organised the advertising on behalf of the Agency, thought that their areas were already sufficiently covered by other newspapers. (In Scotland, for example, the advertisement was run in the Edinburgh *Scotsman* and the *Glasgow Herald*.) After the issue was completed, Morgans analysed how many applications, and how many shares for which applications were accepted, had been produced by each newspaper. The *Daily Telegraph* was far and away the winner.[56]

55 H.C.B.M., Memorandum, 26 May 1953 and K.F.C., 'Steel', 11 June 1953, both File Denationalisation 1953, Box 369, MGP. Bank of England, 'The Sale of British Government Stocks in Payment for Shares of the United Steel Companies Limited', 23 Oct. 1953, File USC 1953, Memos, Box P377, MGP.

56 Altogether, there had been 4,484 applications made on forms from twenty newspapers, and these applicants had received 5.345% of the total number of shares on offer.There had been 1,647 applications made on forms run in the *Daily Telegraph*, 36.73% of all applications made on newspaper forms; the applicants had applied for a total of 272,950 shares, 36.48% of the number of shares accepted on newspaper forms, and 1.95% of the total number of shares on offer. The *Daily Telegraph* was not the biggest circulation paper used, which made its position as winner even more significant. The circulations of both the *Daily Express* and the *Daily Mail* were higher, which was reflected in the charges made for publishing the prospectus: *Daily Telegraph*, £2,840; *Daily Express*, £5,280; *Daily Mail*, £4,576; *The Times*, £1,750; the *Financial Times*, £2,000. The Scottish, Welsh and provincial newspapers were much cheaper: *Manchester Guardian*, £700; Cardiff *Western Mail*, £400; Leeds *Yorkshire Post*, £526; Edinburgh *Scotsman*, £700; *Glasgow Herald*, £800, for example. The total newspaper adver-

Once Morison had decided that United Steel was ready to go, the pace quickened, and by 2 October Morgans had drawn up a timetable for all of the formal arrangements which still had to be completed. On 7 October Erskine met with Morison to settle the final details of the operation. First of all, they decided that, while the prior charges would remain as already decided, the price for the ordinary shares should be 25/- rather than 25/6. No reason was noted down, but November was traditionally a bad month for equities, and Erskine may have insisted on a lower price to ensure success. Secondly, they agreed on the commission to be paid to the Consortium, which would be 6d per share (2%). Out of this would come a commitment commission to the sub-participants (i.e., those who gave irrevocable commitments to take a certain number of shares), and an over-riding commission[57] to brokers of 1/2d per share, leaving 1d per share to be retained by the eight members of the Issuing Houses Consortium and divided equally amongst them. They also agreed that an allotment brokerage of 1 1/2d per share should be paid to the brokers as commission on shares they actually sold to their private clients.[58] This was quite generous. One reason possibly lies in a letter which Erskine wrote to Colville in response to the latter's comments on the May Report: 'It is felt that if we are going to get the banks and brokers generally to take their

tising cost had been £45,314. Charles Barker & Sons, list of newspapers and whether a full page or space for an abridged version was booked, 9 Oct. 1953 and K.F.C., Analysis of newspaper applications, 8 Jan. 1954, both File USC 1953, Memos, and P. Spencer-Smith to Chadwick, 28 Oct. 1953, File USC 1953, all Box P377, MGP. There was a slight contretemps when the *Daily Mail* decided that it wanted to know more than the press reports. In its own version of investigative journalism, the paper sent a journalist to Brown Knight and Truscott Ltd, the printers of the prospectus, to try and find out how many copies they were posting. As it happened, the Works Overseer knew that he should not say anything to anyone, and the only upshot was a stiff letter from the printers to the newspaper. Brown Knight and Truscott Ltd to the Editor of the *Daily Mail*, 26 Oct. 1953, File USC 1953, Box P377, MGP. The total bill from Brown Knight & Truscott Limited for printing, proofing and despatching the various documents in connection with the United Steel sale was £6,828.9.4. Morgan Grenfell to Brown Knight & Truscott, 17 Nov. 1953, File USC 1953, Box P377, MGP.

57 The commission paid to brokers for arranging the sub-underwriting.

58 This would not be paid on irrevocable applications arranged through the Consortium.

coats off and have the considerable work of dealing with applications in the form of Government securities, they will have to be adequately recompensed for their work.'[59]

Morison and Erskine then agreed on the advertising plans, in particular that abbreviated particulars should be placed in a larger number of newspapers than the full prospectus, and on the list of Stock Exchanges where quotations would be requested.[60] All the documents would finally be approved at the following day's meeting of the Consortium, although they would only be finalised on 10 October. There would be a 2% discount if the shares were pre-paid on application; otherwise, they could be paid for in instalments, with 5/- on application, a further 10/- on 15 December 1953 and the final 10/- on 28 January 1954. Although the Agency would subscribe for £10 million 5% debenture stock at 100, and this would be included in the prospectus, the initial payment to United Steel would only be £23 (£2.3 million), with the company able to call the remainder when needed. The announcement to the press would be on 15 October, when a concurrent announcement would be made of the imminence of a sale involving Guest Keen and Vickers. The United Steel shareholders would give formal approval at a meeting of their board of directors on 23 October, and the prospectus would then be published in full on Monday, 26 October 1953.[61]

59 8 May 1953, File Denationalisation, General, Box 369, MGP.

60 Quotation was granted by the London Stock Exchange, as well as by those in Belfast, Birmingham, Bristol, Cardiff, Dublin, Edinburgh, Glasgow, Leeds, Liverpool, Manchester, Newcastle and Sheffield. K.F.C., 'The United Steel Companies Limited', 10 Feb. 1954, File USC 1953, Box P377, MGP. The following provincial brokers were used: G. & W. Beech (Birmingham), B.S. Stock Son & Co. (Bristol), E.T. Lyddon & Sons (Cardiff), Bloxham, Taylor & Martin (Dublin), Bell, Cowan & Co. (Edinburgh), S.M. Penney & Macgeorge (Glasgow), Pilkington & Dunlop (Liverpool), Potter, Stephens & Co. (Leeds), W.A. Arnold & Sons (Manchester), Wise, Speke & Co. (Newcastle) and J.W. Nicholson & Sons (Sheffield). The fees paid to the provincial Stock Exchanges, together with the fee of ten guineas paid to each provincial broker involved, amounted to £1,530.7.6., and the quotation fee of The Stock Exchange, London was £459.7.6. Rowe & Pitman to Morgan Grenfell, 15 Oct. 1953 and Morgan Grenfell to Rowe & Pitman, 10 Nov. 1953, both File USC 1953, Box P377, MGP.

61 K.C. Barrington, 'Matters to be Discussed with Sir John Morison', 7 Oct. 1953, File USC 1953, Docs/Memos, Box P377, MGP. The prospectus was also published on 25 Oct. in the Sunday press and on 30 Oct. in the weekly press.

In the midst of this, the Treasury had second thoughts about the lower issue price Morison had agreed with Erskine. They feared political attack, and Morison hinted as much to Erskine, who wrote a personal letter to reassure him — although the basis of the reassurance was rather worrying:

> You have mentioned to me that the Treasury are a little apprehensive in case there is a runaway market in United Steel shares after dealings begin on the Stock Exchange resulting in a substantial premium over the issue price. We, ourselves, do not believe that there is much likelihood of this happening. In fact, we are more concerned in the opposite direction in case there may be an absence of sufficient buying orders to maintain the price at a premium over the issue price.[62]

Meanwhile, it was time for the brokers to carry out their part in the operation. The six members of the brokers' consortium, Rowe & Pitman, Cazenove, Hoare, W. Greenwell, Panmure Gordon and Joseph Sebag, had met with the Issuing Houses Sub-Committee in May soon after the Report was completed, and they had all received a précis of the Report. Thereafter they had, presumably, been kept informed of developments, in order to talk to the institutions, but they only learned the final decisions as to price after the Issuing Houses themselves had been told at the meeting on 8 October. Thereafter, the brokers polished their official letters to the institutions, which went out on the same day as the press notice, scheduled for 4 p.m. on Thursday, 15 October.[63]

The response of the press was crucial, because it would both create and reflect City reactions to the main points of the sale. (The details would await the publication of the prospectus ten days later.) It was almost uniformly favourable. As the *Financial Times* wrote on 16 October 1953,

> On any comparison with other Heavy Industry equities a yield of 7.2 per cent.[64] with threefold earnings cover would, of course, ensure that the issue would be a roaring success, if it were not for the nationalisation threat. A good 1 per cent. of the yield can be set against that threat and it is up to the individual investor to

62 Erskine to Morison, 13 Oct. 1953, File USC 1953, Box P377, MGP.
63 Meeting of the Consortium of Issuing Houses on Steel, to be held on 8 Oct. 1953, File Denationalisation, Box 369 and S.S. Wilson to Erskine, 12 Oct. 1953, File USC 1953, Box P377, both MGP.
64 Dividend of 9% divided by share price of 25s = yield of 7.2%.

decide whether it is enough.

The 'Lex' column added that 'On whether the margin is adequate views will differ. In new ground such as this, where ordinary criteria for issue terms are insufficient, decisions must be individual, based largely on estimates of the political threat, but my own feeling is that it has been discounted adequately.'[65]

The yield of 7.2% seems to have been the key attraction. The blue-chip average was 5.5%, while that on Industrial Ordinary Index shares was 5.6%,[66] and thus, as the *Financial Times* for 24 October pointed out, 'the United Steel offer does provide an opportunity for increasing the average return on a portfolio.' But other factors were important. First of all, the Account[67] had started well on 14 October, 'with fresh life in the Store and other speculative stocks, a rising trend among the Industrial leaders, and a very firm gilt-edged market.'[68] Secondly, as *The Economist* pointed out on 17 October, 'In United Steel, [the investor] will certainly be buying one of the best equities that the British steel industry has to offer....[It] had made its full contribution to the record outputs that the industry has been winning this year.'[69] *The Economist* then made a further point, which underlined the reason for the earlier insistence of the Issuing Houses on concurrent private sales:

> This operation, and its successors, will still need co-operative support from the institutions...[I]t is clear that the basic ingredient of success will be firm underwriting of the issues and firm holding after allotment.... It cannot fail to be helped, however, by the news that discussions are proceeding between Vickers and Cammell Laird and the Iron and Steel Realisation Agency for the repurchase of their two companies' interests in English Steel and between Guest Keen and Nettlefolds and the Agency on the possibility of a resale of the various steel works taken over from the Guest Keen group.

65 *Financial Times*, 15 Oct. 1953.
66 *Financial Times*, 24 Oct. and 15 Oct. 1953.
67 An Account is a period of two to three weeks within which investors can buy and sell shares without putting up funds or providing stock in settlement.
68 *Financial times*, 15 Oct. 1953.
69 The industry had just announced production of 13,043,000 tons for the first nine months of 1953, 1,204,000 tons above the output of the corresponding period of 1952 and an all-time record. *Financial Times*, 15 Oct. 1953.

While the press was discussing the issue, the brokers were working to complete the underwriting by 22 October. That day, Smith of Morgans wrote to an absent Director that:

> We completed the underwriting of the United Steel issue last night. It has been very well received. We have heard from everybody except about sixteen people who have been offered 90,000 shares. If all these accept we shall have acceptances for about 180,000 shares more than the offer.[70]

The *Financial Times* reported on 22 October that 'The issue has gone extremely well with the institutions. Few failed to take up the amount made available to them. Others wanted more than they had been provisionally allocated. And there were applications by institutions not on the lists.'[71] That sort of response boded well for the flotation.

On Friday, 23 October 1953, the Board of Directors of United Steel met to approve the Offer for Sale and the forms of application, as well as to approve the capital re-organisation. They also went through all the legal formalities necessary to authorise the agreement between the company and the Agency, to create the debenture stock, to receive payment for the debentures and preference shares and to approve the capitalisation of reserves.

70 Smith to Viscount Harcourt, 22 Oct. 1953, File USC 1953, Box P377, MGP.

Commitments by 21 Oct. 1953

	Acceptances	Refunds	Outstanding	Total Amounts Offered
Rowe & Pitman	7,038,000	522,000	44,000	7,604,000
Cazenove	6,068,500	266,000	46,000	6,380,500
W. Greenwell	108,000	7,000	-	115,000
Hoare	225,000	-	-	225,000
Panmure Gordon	50,000	-	-	50,000
Joseph Sebag	175,000	-	-	175,000
	13,664,500	795,000	90,000	14,549,500

71 Six of the Issuing Houses in the Consortium also made irrevocable applications for shares (and received $4^1/_2$d per share thereby): Barings (100,000 shares), Bensons (20,000), Helbert Wagg (30,000), Lazards (20,000), Morgan Grenfell (140,000) and Schröders (40,000). Rothschilds and Hambros stood out. Irrevocable applications for the other 13,650,000 shares were obtained by the consortium of brokers. K.F.C., 'The United Steel Companies Limited', 10 Feb. 1954, File USC 1953, Box P377, MGP.

The Company then handed a cheque to the Agency to pay off the loan and the Agency formally approved the Offer for Sale. Over at the Bank, tender prices were being fixed for all of the eligible Government stocks.[72] On Saturday the Offer for Sale plus pink forms were posted to former shareholders, on Sunday the prospectus was published in eight Sunday papers, and on Monday, 26 October 1953, the prospectus was printed in thirty national and regional newspapers. (The four weeklies published it on 30 October.) The lists were then open until 9 November.

The Outcome of the Flotation

The result was impressive: as Bicester wrote to the Governor on 16 November, 'It has all been most satisfactory, but I do not think that any of us quite anticipated the success we have met with.'[73] About 52,400 applications were received for a total of approximately 40,389,000 shares, so that the issue was over-subscribed three times. Applications for nearly 61% of the 14 million shares on offer were from former shareholders, and Government stocks were accepted by the Agency for approximately 21 1/2% of the shares. The scale of the response meant that applications had to be scaled down. Former shareholders received favourable treatment: applicants for ten thousand or fewer shares received all they wanted, while larger applications received 23%, with a minimum of ten thousand. Public applicants for two hundred or fewer shares received what they applied for, while larger applications received about 11%, with a minimum of two hundred.[74]

72 United Steel Companies, Minutes of the 311th meeting of the Board of Directors, 23 Oct. 1953, and 'The United Steel Companies Limited Time Table', 2 Oct. 1953, both File USC 1953, Box P377, MGP.

73 File USC 1953, Box P377, MGP.

74 About 8,600 applications for 8,504,000 shares were from former shareholders; Government stocks were accepted for approximately 3 million shares. K.F.C., 'The United Steel Companies Limited', 10 Feb. 1954, File USC 1953, Box P377, MGP. Morison wrote to Erskine on 13 Nov. 1953 that 'Out of total tenders of Government Stock of 10,000 in number, 5,500 were on ordinary forms and 4,500 on "pink" forms. The 5,500 ordinary forms included 2,000 offering Steel Stock, 1,500 offering 3 1/2% War Loan, and the balance covered a miscellaneous selection. Of the 4,500 "pink" form applications, 4,000 represented Steel Stock.

Dealing in the shares of United Steel began on Monday, 16 November, with 'a big two-way turnover...involving between 2m. and 3m. shares'. The market 'was completely free of "shop" intervention, yet the premium varied by only 1$^1/_2$d throughout the day, closing at 3d over the issue price of 25s. The consortium were reported as dealing, both for stags and for fresh buyers, but there were no "peg" prices.'[75] By the end of the week, however, the premium 'had almost entirely run off. The first day's volume of business was brisk,...but nothing of interest [had] developed since.' The disturbance caused by an engineering strike was alleged to be a 'hampering influence', as well as the need of 'the investing public...to keep its powder dry for the flotation still to come.'[76] For whatever reason, the aftermarket did not hold up, and by 20 January 1954 the discount on the offer price was 1/3 [7$^1/_2$p],[77] or 6%.

Why did the shares fall to a discount, where they remained for nearly a year? The argument that 'By far the most important point was the fear of renationalisation and the terms of compensation'[78] was taken for granted, but the manner in which this fear affected the behaviour of investors was open to debate. Morison analysed the reaction of the institutions in a mood more of sorrow than anger:

> It was always clear that such a vast operation...would require the whole-hearted co-operation of the institutional investor. It was essential that the Institutions should be prepared to take and hold for a considerable time much larger blocks of Steel shares than they would want as permanent holdings. The transfer of the shares from the Institutions into the hands of other investors was a matter that would take a very long time to effect.
>
> The initial co-operation of the Institutions was excellent, but they have taken a jaundiced view of the pronounced lack of enthusiasm shown by the large private concerns, who previously had substantial Steel interests, to re-purchase these interests. These private concerns have not measured up to the standard required of them

It is interesting to see what a large proportion of the previous Shareholders who tendered Stock have offered Steel Stock in payment.' File USC 1953, Box P377, MGP. Government Stock could not be tendered in the 1961 sale of prior charge securities.

75 *Financial Times*, 17 Nov. 1953.

76 *Financial Times*, 21 Nov. 1953.

77 *The Scotsman* 20 Jan. 1954.

78 Ralph Vickers of Vickers, da Costa & Co., to E. de Rothschild, 5 Feb. 1954, File Denationalisation, General, Box 369, MGP.

by the Institutions. In the result, the Institutions have not been enthusiastic in supporting the market in United Steel shares and adequate "after care" of the market has been lacking.

Morison went on to add that 'the principal trouble is the large amount involved', which under the most favourable conditions could result in 'market indigestion' (particularly after the second flotation, that of Lancashire Steel, in January 1954).[79] Support for this view came from the Managing Director of the Anglo-Scottish Amalgamated Corporation Limited, a sub-participant in the issue, who wrote that 'support is not available primarily because the institutions have taken firm commitments for as much of the stock as they want to hold.'[80] For this situation, according to Ralph Vickers of Vickers, da Costa & Co., the Consortium was to blame for 'The underwriting method adopted, which ensured that a lot of stock was issued to Institutions and individuals who had no wish to hold the stock, and which therefore made certain that there would be substantial sellers when dealings commenced.'[81]

This was clearly true, and indeed, the institutions had made their intentions clear. But Morison apparently did not blame the Issuing Houses for this outcome, largely because he had agreed that there was no better alternative to the approach which had been taken. It had been a gamble that there existed an investing public with a yen once again to hold steel shares, but as Erskine wrote to Morison on 10 February 1954, 'the state of the "after market" in United Steel shares...indicated that although

79 Memorandum by Morison, 23 March 1954, File Denationalisation, Docs/Memos, Box 369, MGP. Schröders led the Issuing Houses Consortium for the Lancashire Steel flotation.

80 Douglas Hewitt to Panmure Gordon & Company, 26 Jan. 1954, File Denationalisation, General, Box 369, MGP. As Bicester noted to the Morgan Grenfell shareholders on 15 June 1954, 'in subsequent market dealings the shares fell below the issue price due mainly to an absence of buying orders from institutional investors, very few of whom "made up" their comparatively small allotment.' Chairman's Speech for 1953, the Secretariat, Morgan Grenfell & Co. (hereafter MGC).

81 Vickers to Rothschild, 5 Feb. 1954, File Denationalisation, General, Box 369, MGP. H. Ockford, Managing Director of the Industrial and General Trust, Limited, also complained in early February 1954 that the share price had borne no relation to market conditions; this, however, seems dictated by hindsight, since he could have declined participation at the time. Ockford to Panmure Gordon & Co., 2 Feb. 1954, File Denationalisation, General, Box 369, MGP.

the public generally might come in for a quick turn, they did not appear at the present time to be permanent investors in any large volume in steel equities.'[82] By early March, the largest shareholder in United Steel was the Co-Operative Insurance Society of Manchester, which held over 300,000 shares, while the next three largest, with between 100,001 and 200,000 each, were Morgan Nominees, London and Manchester Assurance, and the Prudential. There were very few individual investors with more than 10,000 shares, while the average holding below that level was 357 shares.[83] The latter number, however, implies 29,742 small investors, so presumably Erskine was lamenting the number of shares held by the average shareholder, not the number of shareholders themselves. Indeed, one result of the several denationalisation issues would be that steel company shares were more widely distributed than those of companies generally.[84]

What had been the financial recompense of the Issuing Houses and others for their labour in the operation? The Consortium received a commission of 6d per share on 14 million shares, for a total of £350,000. Out of this they had to pay the commitment commission of 4 1/2d per share to themselves, institutions, stock brokers and others who had made irrevocable applications for shares. The commitment commission came to £262,500. Next, there was the overriding commission of 1/2d per share, which came to £29,166.13.4, which was also subtracted from the £350,000. Finally, there was the legal fees paid by the Consortium to Slaughter and May for advice on the United Steel offer, which came to £4,200. Altogether, these came to £295,866.14.4, which left £54,133.13.4, from which were subtracted the legal fees paid to Linklaters & Paines and to Slaughter and May for advice on the sale of steel shares by the Agency, which came to £3,150, and the Counsel's fee of £30.15.6. The final profit for the Consortium, then, came to £50,952.11.2 which, when divided eight ways, gave each Issuing House a profit of £6,369.1.5. Morgan

82 File Denationalisation, General, Box 369, MGP.

83 *Financial Times*, 5 March 1954. The remaining large holdings between 10,000 and 100,000 shares were mainly held by other institutional investors, such as the insurers Pearl Assurance and Eagle Star, and by a number of pension schemes, such as that of Unilever. There were 117 holders with between 10,000 and 50,000 shares, and 10 holders with between 50,001 and 100,000. The 131 holders of over 10,000 shares represented an aggregate of 3,382,000 shares. *Ibid.*

84 B.S. Keeling and A.E.N. Wright, *The Development of the Modern British Steel Industry* (London: Longmans, Green, 1964), p.179.

Grenfell in addition received from United Steel a fee of £10,000 for advice on the capital reconstruction of the company.[85] The Governor had asked that, for political reasons, the Consortium's fee be kept down,[86] and in this respect, the Government could certainly have no complaints.

Indeed, the Government had reason for satisfaction on every level. First of all, in the structure and membership of the Iron and Steel Board it had its preferred organisation and mostly its preferred membership in place to supervise the industry, allowing the Board of Trade happily to devolve responsibility onto the Board. As Sandys wrote to Forbes in December 1953, 'I need hardly say how glad I am that you are now relieving me of this awkward child.'[87] The fact that the Board lacked the powers to impose painful but arguably necessary solutions to the industry's problems would only become obvious over the subsequent fourteen years. Secondly, the resale of United Steel had been a very great success and thus boded well for the whole programme, in addition to which the Government had made a profit on the transaction. And finally, the Government had achieved a larger political end: by returning United Steel to the private investor, they had struck a blow for the principle of free enterprise, at the same time setting new boundaries for the proper role of the State. It was as well for the Government's self-satisfaction that they did not then realise how moveable those boundaries would prove to be.

85 K.F.C., 'The United Steel Companies Limited', 10 Feb. 1954, and Memo from S.S. Wilson (of the Agency), 3 Dec. 1954, both File USC 1953, Box P377, MGP.

86 'We wholly concur with the Governor's request that care should be taken to avoid any accusation that "the City" is making undue profit and we are of the opinion that the percentage spread chargeable to the Agency should be kept as low as possible.' 'Report by the Sub-Committee on Steel', 13 May 1953, File Denationalisation, Docs/Memos, Box 369, MGP.

87 Sandys to Forbes, 7 Dec. 1953, BT 255/123, No.E69.

5 THE EPILOGUE

The period between 1953 and 1967 saw the gradual denationalisation of most of the steel industry and then its abrupt renationalisation. By January 1955 the equity of six of the largest companies had been resold to the private sector, followed by that of several of the smaller companies and two of the larger ones over the following two years, as well as some of the debt. Towards the end of 1957, however, the share market weakened, and it was only in 1961 that the Agency was able to sell most of its large holdings of prior charge securities. The general election of October 1964 brought the Labour Party to power, and the Government then set in motion the preparations for the renationalisation of the industry. The Iron and Steel Act 1967 was the result, and this established the British Steel Corporation. Changes in its corporate structure two years later finally terminated the separate existence of the companies producing bulk steel.

The euphoria generated by the immediate results of the United Steel flotation encouraged the sale of Lancashire Steel Corporation Ltd in January 1954,[1] but in this case the lack of response by the investing public ensured that the outcome was very different, and the share prices of both United Steel and Lancashire Steel subsequently fell to substantial discounts. This precluded any more flotations for the time being, and the result was some disquiet in the Conservative constituencies.[2] But the market outlook changed, and the turning point was reached in June 1954

1 The offer was for 4.5m Ordinary Shares of £1 each @ 22/- and 4.5m 5 3/ 4/% Redeemable Cumulative Preference Shares of £1 each @ 21/-. The Consortium was led by Schröders, and each member made a profit of £3,253.2.6. Chairman's Speech for 1954, the Secretariat, MGP.

2 Lord Woolton wrote to the Prime Minister on 22 March 1954 that 'As a result of conversations I have had recently with people who have been in London for the Party meetings, I find some concern in the Party in the Country as to the extent to which our denationalisation programme is effective. We have passed the Acts, but the question they are asking is to what extent the denationalisation of... iron and steel is, in fact, taking place.' Ms 22, f.93, Woolton Papers, Bodleian Library, Oxford.

when the shares of Stewarts & Lloyds, the most profitable of the companies, were successfully sold.[3] Two further issues in 1954, those of John Summers & Sons Ltd[4] and Dorman, Long & Co. Ltd,[5] met with increased success, and the growing momentum culminated in the 'overwhelming response'[6] in January 1955 to the Colvilles Limited offer of 10 million shares: 150,000 applications were received for a total of 130,000,000 shares.[7]

> With the successful completion of this stage, the Issuing Houses agreed that in the case of the smaller steel companies the machinery of the Consortium would be unnecessarily elaborate. The Consortium was therefore dissolved on the understanding that each House should separately underwrite the Offers for Sale in the case of the companies for whom they were accustomed to act and the equity capital (and in two cases the loan capital as well) of the following companies was successfully sold to the investing public during 1955:[8] [Whitehead Iron and Steel Co.,[9] Thos. Firth & John Brown Ltd,[10] Hatfields[11] and Consett Iron Company.[12]]

The arrangements for the Whitehead offer were largely complete before the Consortium was dissolved, and Barings (who under-

3 The offer was for 10m Ordinary Shares of £1 each @ 35/-. The Consortium, each member of which made a profit of £2,625, was jointly led by Morgan Grenfell and Helbert, Wagg, who shared an additional fee of £5,250 for advising the company. Chairman's Speech for 1954, the Secretariat, MGC.
4 The offer was for 9m Ordinary Shares of £1 each @ 24/6. The Consortium, each member of which made a profit of £4,194.15.10, was jointly led by Morgan Grenfell and Helbert, Wagg, who shared an additional fee of £10,000 for advising the company. Chairman's Speech for 1954, the Secretariat, MGC.
5 The offer was for 15m Ordinary Shares of £1 each @ 22/6. The Consortium, each member of which made a profit of £7,518.4.7, was jointly led by Morgan Grenfell and Lazards, who shared an additional fee of £10,000 for advising the company.
6 Chairman's Speech for 1954, the Secretariat, MGC.
7 The offer was for 10m Ordinary Shares of £1 each @ 26/-. The Consortium, each member of which made a profit of £4,876.16.8, was led by Morgan Grenfell, who received an additional fee of £10,000 for advising the company. Chairman's Speech for 1955, the Secretariat, MGC.
8 Chairman's Speech for 1955, the Secretariat, MGC.
9 Underwritten by Barings.
10 Underwritten by Hambros.
11 Underwritten by Lazards.
12 Underwritten by Hambros.

wrote the offer) decided to share their profit with the other members of the Consortium.[13]

In view of the progress made with the sale of equity capital, the Agency decided to begin selling the prior charge capital which it was holding. In May 1955 Morgan Grenfell successfully underwrote the offer for sale of the preference shares and debenture stock of the English Steel Corporation,[14] while in 1956 they underwrote the sale of both the ordinary shares and debenture stock of South Durham Steel and Iron Company Limited.[15] The climax to this stage in the denationalisation of the industry was reached in 1957 when the Consortium underwrote the offer for sale of 40 million ordinary shares of The Steel Company of Wales Limited, 'the largest equity issue ever made in the City of London'.[16] Between 1953 and 1957, 86% of the productive capacity of the iron and steel industry had been denationalised.[17]

Thereafter the programme ran into difficulties. The largest single entity which remained was Richard Thomas & Baldwin, which was known to be considering a £100 million steel works and strip mill development programme; as a result, it would be impossible for some years for investors to check its earning power against actual results, and in fact, it was never sold. As for most of the others, they included plant which was old or badly located or both. But even if the state of the companies had been more encouraging, it might have been difficult to sell them, since the equity market weakened towards the end of 1957 and would therefore have been less welcoming for new issues. The death of Morison in March 1958 was a further blow.[18]

By 1960, however, the investment climate had again changed, and Morgan Grenfell and the Agency began to plan the sale of the Agency's holdings of prior charge securities. In November

13 Each member received £1,246.1.10. Chairman's Speech for 1955, the Secretariat, MGC.

14 The offer was for 5m 5 1/ 2% Cumulative Redeemable Preference Shares of £1 each @ 20/- and £5m 4 1/ 2% Debenture Stock 1974/79 at par. Their profit was £40,321.17.3. Chairman's Speech for 1955, the Secretariat, MGC.

15 The offer was for 8m Ordinary Shares of £1 at 27/6 and £3m 5 1/ 2% Debenture Stock 1976/81 at 97%. Morgan's profit was £42,236.9.2. Chairman's Speech for 1956, the Secretariat, MGC.

16 Chairman's Speech for 1957, the Secretariat, MGC. Each member of the Consortium made a profit of £18,282.3.4. *Ibid*.

17 Keeling and Wright, *British Steel Industry*, p.179.

18 *Ibid*.

1960, Erskine suggested a 'package-deal' operation,[19] and, after discussions amongst the Agency, the Treasury, seven of the denationalised steel companies and Morgan Grenfell, 24 February 1961 saw the simultaneous publication of seven prospectuses offering for sale twelve separate securities held by the Agency.[20] (Two others were close to maturity and the Treasury reluctantly agreed not to include them in the block sale, while another was deemed unmarketable.)[21] Finally, 1963 saw the last denationalisation sales, of Barrow Iron Works, Gjers Mills, and the remaining steel companies (except for Richard Thomas & Baldwin). It is worth noting that for assets with a total book value of £374 million, the Agency received £388 million.[22]

A general election had been held on 8 October 1959, and the

19 Herbert Brittain of the Agency to Erskine, 22 Dec. 1960, File 1961 Offer for Sale of the... Agency Prior Charges, Box 362, MGP.

20 These were (1) Colvilles' 4m 5 1/ 2% Cumulative Preference Shares; (2) Colvilles' £9,820,000 4 1/ 2% Debenture Stock 1975/85; (3) Consett's £3,924,150 5% Redeemable Debenture Stock 1975/85; (4) Dorman, Long's 5m 5 1/ 2% Cumulative Preference Shares; (5) Dorman, Long's £10m 4 1/ 2% Unsecured Loan Stock 1969/74; (6) John Summers' 4m 5 1/ 2% Cumulative Preference Shares; (7) Steel Company of Wales' £40m 5 1/ 2% First Debenture Stock 1980/85; (8) Stewarts & Lloyds' 5m 5 1/ 2% Cumulative Preference Shares; (9) Stewarts & Lloyds' 5m 5 1/ 2% Redeemable Cumulative Preference Shares; (10) United Steel's 5,097,000 4 1/ 2% Cumulative Preference Shares; (11) United Steel's 4m 5 3/ 4% Redeemable Cumulative Preference Shares; and (12) United Steel's £9,600,000 4 3/ 4% Debenture Stock 1968/78. Freshfields to The Stock Exchange, draft of 13 Jan. 1961, File 1961 Offer, Box 362, MGP. Morgan Grenfell received a fee of £70,000, Lazards, Helbert Wagg and Hambros each received £40,000, and Barings, Bensons, Rothschilds and Schröders each received £26,250. Horsfall to Erskine, 21 Feb. 1961, File 1961 Offer, Box 362, MGP.

21 The three were (1) John Summers' £8m 4 1/ 2% Unsecured Loan Stock 1962-64, (2) Stewarts & Lloyds' £10m 4 1/ 2% Unsecured Loan Stock 1964 and (3) Steel Company of Wales' £25m 5 3/ 4% Second Debenture Stock 1964-87. 'The Treasury accept of course the advice which you have given about unsuitability of the first two stocks for inclusion in a block offer and the unmarketability of the third stock as it stands. On the other hand they say that they must have regard to the heavy impending calls on the Realisation Account and the Exchequer in respect of R.T.B. and Colville requirements and the desirability of ensuring that as far as practicable these are covered by accruing proceeds from sales.' Brittain to Erskine, 31 Oct. 1960, File 1961 Offer, Box 362, MGP.

22 Keeling and Wright, *British Steel Industry*, pp.179-80.

next therefore had to be held by October 1964.[23] What was uncertain was the outcome, which was reflected in the volatility of steel company shares (as had been the case before the 1959 election).[24] It was clear what would happen if Labour won:

> In every debate on the industry in Westminster from 1951 onwards, at every party conference, in every election manifesto, it was an article of faith in the Labour creed to stress the certainty of steel renationalisation when and if the party returned to office. Correctly or incorrectly, many Labour politicians and supporters saw steel nationalisation as the issue that had first loosened and then broken their grip on office after the war....[25]

Labour did win, but with a majority of only four, and although the decision to renationalise was probably taken early in 1965, on 18 March 1965 the Cabinet decided to delay introducing a bill, and instead to publish a White Paper containing the proposals being drawn up by the Ministry of Power. There was little pressure in favour of renationalisation from the steel industry unions, and some opposition to it from elements within the party who believed that they should concentrate on urgent social measures. The Labour Prime Minister, Harold Wilson, took a tactical decision to stall until another election had increased Labour's majority.[26]

The White Paper was published on 30 April 1965, and it included the following major points: (1) the thirteen principal bulk producers of iron and steel were to be taken over; (2) substantial sections of the 1949 Bill would be resuscitated; (3) the nationalised company was to be under the control of a board of directors which would be appointed by the Minister of Power and responsible to him; (4) compensation was to be paid at a rate determined by the average value of shares over a five-year period before 1964; and (5) unlike the previous bill, the structure of the new corporation would not be defined (and therefore made rigid) by legislation. As one political scientist has noted, 'Conser-

23 Indeed, it was a week late — 15 Oct. 1964. The previous election had been held on 8 Oct. 1959.

24 Richard Roberts has demonstrated that steel shares, as reflected in the FT-Actuaries Steel Index, both fell faster and rose faster than the market as a whole as represented by the FT-30 Index. Roberts, 'Election Values', *Financial Times*, 30 May 1987.

25 Keith Ovenden, *the Politics of Steel* (London: The Macmillan Press Ltd, 1970), p.26.

26 *Ibid*, pp.45-47.

vative antagonism to the measure was stifled somewhat by the high level of compensation payments, so that on the morning of May Day, ironically enough, shareholders in the steel companies that were to be nationalised found themselves much richer than they had been on the day before.' Certainly, steel share prices rose rapidly.[27]

A general election on 31 March 1966 provided Labour with a majority of ninety-six, and on 29 June 1966 the Government published its Bill. It contained substantially the same proposals as the White Paper, except that compensation had been lowered by £71 million to £484 million, 'sparking off the depressing but predictable after-hours dealing in Throgmorton Street.'[28] For Richard Marsh, the Minister of Power, as for Duncan Sandys, getting the Bill through a Cabinet Committee had proved difficult. In Marsh's case, both George Brown, the Secretary of State for Economic Affairs and later the Foreign Secretary, and James Callaghan, the Chancellor of the Exchequer, had opposed it. Callaghan's opposition had probably reflected the Treasury view, which was opposed to renationalisation and the attendant compensation payments at a time of continuing economic difficulty; they were eventually mollified by an amendment.[29]

The Bill had a somewhat rocky Commons passage: the Standing Committee sessions, which began on 25 October 1966, consumed thirty-one sittings, including one of forty-four hours and thirty-seven minutes, then the longest session for a Standing Committee in the history of Parliament.[30] Nevertheless, it received its Third Reading on 26 January 1967 with a majority of eighty-six. Its passage in the Lords was 'swift and uninterrupted': the fact that the Bill had been introduced early in the life of the Parliament 'effectively silence[d] opposition from the Lords and discourage[d] them from blocking its passage.'[31] On 22 March 1967 the Iron and Steel Act 1967 received the Royal Assent.

The new British Steel Corporation mirrored the structure of the 1949 version, in that the Corporation was a holding com-

27 P.P. 1964-65, Cmnd. 2651. Ovenden, *Politics of Steel*, p.48, for the quotation. Colvilles went up 14/- to 42/3, South Durham up 9/6 to 28/4, Lancashire Steel and United Steel both up 8/1. Ovenden, pp.217-18.
28 Ovenden, *Politics of Steel*, pp.106-7.
29 *Ibid*, pp.107-8.
30 *Ibid*, p.110. This record was surpassed in 1968 by the Transport Bill.
31 *Ibid*, p.111.

pany with the fourteen bulk steel producers (the previously denationalised thirteen plus Richard Thomas & Baldwin) as its subsidiaries. But by February 1969 changes were introduced, the point of which was to move from an initial structure organised primarily on a regional basis to one organised on a product basis. On the one hand, this was a development which had been mooted as early as 1951 under the old Iron and Steel Corporation. On the other, it was hastened by the fact that the United Steel Companies and Stewarts & Lloyds, represented on the Corporation's Board by A.J. Peach and Niall Macdiarmid, appeared reluctant to co-operate with the Corporation,[32] presumably in the hope that history would again repeat itself. Structural reorganisation of the Corporation on a product basis obliterated the old company structures. This had the additional effect of rendering a future straight re-denationalisation considerably more unlikely, which in any case the Conservative leadership were very cagey about promising, making no reference to it from the beginning of the Committee Stage of the 1967 Bill onwards. For the next ten years motions urging denationalisation were never selected for debate at Conservative Party Conferences.[33]

The Conservatives believed that the old Iron and Steel Board had not been a success. This was a perception which was widely shared, not least by the Board itself, which complained about its lack of powers in its thirteenth and last *Report* on 5 July 1967. *The Economist* was not very sympathetic, pointing out that

> the board itself cannot escape criticism: it was always cautious and conservative and if the members were aware of the weaknesses in its powers as they now say, they should have made more of a song and dance about righting them during the board's lifetime rather than lamenting them at its funeral.
>
> Certainly the board's powers were always too limited and of the wrong kind.... In effect, therefore, the board had some power to tell the industry what not to do; it had none to tell the steel masters what to do. Instead of acting as an agency to promote closures, mergers and rationalisation, the board has been under political pressures to encourage the setting-up of new plants in development areas, irrespective of whether the siting was the best for the industry's efficiency. ...The inefficient survived.[34]

The argument that the Board had no positive powers of control

32 Private information.
33 Ovenden, *Politics of Steel*, p.124.
34 *The Economist*, 8 July 1967.

had of course been made by George Strauss during the debates in 1952 and 1953, and it might have been thought that the renationalisation of the industry would have his strong support. This was not the case. By 1965 Strauss believed that the Government should set up a board with strong powers of control, but that there should be no element of public ownership.[35] This was a position with which some members of the Conservative Party could sympathise, but in the end the British Steel Corporation remained in being. The steel industry was believed by too many in the Conservative Party leadership, Whitehall and the press to have performed inadequately while under private ownership. But even if the political will once again to denationalise iron and steel had been present, the City would have needed to be convinced of its merits, and certainly at least one institution refused even to consider the possibility of investing in denationalised steel shares once again.[36] It would take a generation whose experience of steel stemmed primarily from its performance as part of the public sector to plan once again to set the industry free.

35 Ovenden, *Politics of Steel*, p.61.

36 C.G. Randolph, Sun Life Assurance, wrote to J.E.C. Collins, Morgan Grenfell, on 12 July 1966 that '...we are very luke-warm about any suggestion being made of the Conservatives' intention to denationalise Steel in the future. The Steel industry has been a political pawn for far too long and as far as the long term investor is concerned, the time must now have arrived (with the announcement of the revised renationalisation terms) when this section of the market should be disregarded by him. There seems to be little point investment-wise for the politicians to attempt to unscramble the position yet again, for not only is the industry likely to remain on a lengthy cyclical downturn for some years to come, but with repeated uncertainty of his future position, it is considered that the Institutional investor will be unwilling to invest in most of the individual stocks which would be resuscitated a second time.' File 1953 Denationalisation, Box 369, MGP.

DRAMATIS PERSONAE

BICESTER, *see* V.H. Smith

BUTLER, R.A., 1902–82. Baron Butler, cr. 1965 (Life Peer). Marlborough and Pembroke College, Cambridge. MP (C) for Saffron Walden 1929–65. Under-Secretary of State, India Office, 1932–37; Parliamentary Secretary, Ministry of Labour, 1937–38; Under-Secretary of State for Foreign Affairs 1938–41; Minister of Education 1941–45; Chairman, Conservative Research Department, 1945–64; Chancellor of the Exchequer 1951–55; Leader of the House of Commons 1955–61; Lord Privy Seal 1955–59; Home Secretary 1957–62; First Secretary of State and Deputy Prime Minister July 1962–October 1963; Secretary of State for Foreign Affairs 1963–64. Master of Trinity College, Cambridge 1965–78.

CECIL, Robert Arthur James Gascoyne-, 1893–1972. 5th Marquess of Salisbury, cr. 1789, succeeded 1941. Eton and Christ Church, Oxford. MP (C) for South Dorset 1929–41. Parliamentary Under-Secretary of State for Foreign Affairs 1935–38; Paymaster General 1940; Secretary of State for Dominion Affairs 1940–42, 1943–45; Secretary of State for the Colonies 1942; Lord Privy Seal 1942–43, October 1951–May 1952; Secretary of State for Commonwealth Relations May–November 1952; Lord President of the Council 1952–57; Acting Secretary of State for Foreign Affairs June–October 1953.

COBBOLD, Cameron Fremanteel, 1904–88. 1st Baron, cr. 1960. Eton and King's College, Cambridge. Bank of England 1933–61; Executive Director, 1938–45; Deputy Governor 1945–49; Governor 1949–61. Lord Chamberlain of HM Household 1963–71.

CROOKSHANK, Harry Frederick Comfort, 1893–1961. 1st Viscount Crookshank, cr. 1956. Eton and Magdalen College, Oxford. Military service 1914–19. Diplomatic Service 1919–24. MP (C) for Gainsborough 1924–56. Parliamentary Under-Secretary of State for Home Affairs 1934–35; Parliamentary Secretary for Mines 1935–39; Financial Secretary to the Treasury 1939–43; Postmaster-General 1942–1945; Minister of Health 1951–52; Leader of the House of Commons 1951–55; Lord Privy Seal 1952–55.

DUNCAN, Andrew Rae, 1884–1952. Knight, cr. 1921. Glasgow University. Barrister. Coal Controller, 1919–20. Vice-President, Shipbuilding Employers' Federation, 1920–27. Chairman, Central Electricity Board, 1927–35; Director, Bank of England, 1929–40. Chairman, Executive Committee, British Iron & Steel Federation, 1935–40, 1945–52. MP (Nat.) for City of London 1940–50. President of Board of Trade, 1940, 1941–42; Minister of Supply, 1940–41, 1942–45.

ERSKINE, (Robert) George, 1896–1984. Knight, cr. 1948. Kirkcudbright Academy and Edinburgh University. National Bank of Scotland 1913–

29; Morgan Grenfell 1929–67, Managing Director, 1945–67. Deputy-Chairman of NAAFI 1941–52. Chairman, London & Provincial Trust Ltd 1954–71.

EVANS, Lincoln, 1889–1970. Knight, cr. 1953. Swansea Elementary School. General Secretary, the Iron and Steel Trades Confederation, 1946–53. Member of Iron and Steel Board 1946–48; Member of Economic Planning Board 1949–53; Deputy Chairman, British Productivity Council, 1952–53; Deputy Chairman, Iron and Steel Board, full-time 1953–60, part-time 1960–70.

FIGGURES, Frank Edward, 1910-. Knight, cr. 1970. Rutlish School and New College, Oxford. Called to Bar, Lincoln's Inn, 1936. Military service (RA) 1940–46. Joined HM Treasury 1946; Director of Trade and Finance, OEEC, 1948-51; Under-Secretary, Treasury, 1955-60; Secretary General to EFTA 1960–65; Third Secretary to Treasury 1965–68; Second Permanent Secretary, Treasury, 1968–71; Director-General, National Economic Development Council, 1971–73; Chairman, Pay Board, 1973–74.

FORBES, Archibald Finlayson, 1903-. Knight, cr. 1943. Paisley and Glasgow University. Chartered Accountant, member of Thomson McLintock & Co. until 1935. Executive Director, Spillers Ltd, 1935–65, Chairman, 1965–68, President, 1969–80. Deputy Secretary, Ministry of Aircraft Production, 1940–43; Controller of Repair, Equipment and Overseas Supplies 1943–45. Chairman, first Iron and Steel Board, 1946–49; Chairman, Iron and Steel Board, 1953–59. Chairman, Debenture Corp. Ltd, 1949–79, and other companies; Chairman, Midland and International Banks Ltd, 1964–76, President, Midland Bank Ltd, 1975–83.

GILBERT, Bernard William, 1891–1957. Knight, cr. 1946. Nottingham High School and St John's College, Cambridge. Joined Treasury 1914; Joint Second Secretary to the Treasury 1944–56.

HUNTER, Ellis, 1892–1961. Knight, cr. 1948. Educated privately. Chartered Accountant. Partner, W.B. Peat & Co. (now Peat Marwick McLintock). Managing Director, 1938–61 and Chairman, 1948–61, Dorman, Long & Co. Ltd. President, British Iron & Steel Federation, 1945–53.

LINDSELL, Herbert George, 1903–73. Educated at elementary and secondary schools. Board of Trade 1918–32; Import Duties Advisory COmmittee 1932–39; Ministry of Supply 1939–59, Under-Secretary 1950–59. Principal Officer (Establishments), United Kingdom Atomic Energy Authority, 1959–63.

MORISON, John, 1893–1958. Knight, cr. 1945. Greenock Academy. Chartered Accountant; partner in Thomson McLintock & Co. (now Peat Marwick McLintock), 1926–58. Director-General (Finance and Contracts), Ministry of Supply, 1942–45. Chairman, Iron and Steel Holding and Realisation Agency, 1953–58.

MYNORS, Humphrey Charles Baskerville, 1903-. 1st Baronet, cr. 1964. Marlborough and Corpus Christi College, Cambridge. Fellow of Corpus 1926–33. Bank of England 1933–64; Executive Director of the Bank 1949–54; Deputy Governor 1954–64. Chairman, Panel on Take-overs and Mergers, 1968–69, Deputy Chairman 1969–70.

PEACOCK, Edward Robert, 1871–1962. Knight, cr. 1934, for services to the Royal Family. Queen's University, Kingston, Canada. English master and senior house master, Upper Canada College, Toronto, 1895–1902. Dominion Securities Corporation of Canada and London 1902–15. Company doctor for public utility companies in Spain, Brazil and Mexico 1915–24. Managing Director of Baring Brothers 1924–54. Director of Bank of England 1921–24, 1929–46. Receiver-General of Duchy of Cornwall 1929–54.

PHILLIMORE, John Gore, 1908-. Winchester College and Christ Church, Oxford. Partner of Roberts, Meynell & Co., Buenos Aires, 1936–48. Representative of HM Treasury and Bank of England in South America 1940–45. Managing Director of Baring Brothers 1949–72.

SALISBURY, *see* Cecil.

SANDYS, Duncan Edwin, 1908–87. Baron Duncan-Sandys, cr. 1974 (Life Peer). Eton and Magdalen College, Oxford. Married Diana Churchill, 1935 (divorced 1960). Diplomatic Service 1930–35. MP (C) for Lambeth Norwood 1935–45 and for Streatham 1950–Feb. 1974. Financial Secretary to the War Office 1941–43; Parliamentary Secretary, Ministry of Supply, 1943–44; Ministry of Works 1944–45; Minister of Supply 1951–54; Minister of Housing and Local Government 1954–57; Minister of Defence 1957–59; Minister of Aviation 1959–60; Secretary of State for Commonwealth Relations 1960–64 and Secretary of State for the Colonies 1962–64. Chairman, Lonrho Ltd, 1972–84, President 1984–87.

SMITH, Randall Hugh Vivian, 1898–1968. 2nd Baron Bicester, cr. 1938. Eton and Sandhurst. Managing Director, Morgan Grenfell & Co. Ltd, 1938–68.

SMITH, Vivian Hugh, 1867–1956. 1st Baron Bicester, cr. 1938. Eton and Trinity Hall, Cambridge. Hay's Wharf (the family firm) c. 1889–1905. Managing Director, Morgan Grenfell & Co. Ltd, 1905–56, senior Managing Director 1941–56. Governor, Royal Exchange Assurance Corporation, 1914–56. For many years Chairman of the Conservative and Unionist Association of the City of London.

STRAUSS, George Russell, 1901-. Baron Strauss, cr. 1979. Rugby.
MP(Lab) for Lambeth North 1929–31, 1934–50; MP (Lab) for Lambeth Vauxhall 1950–79. Parliamentary Secretary to Minister of Transport 1929–31; Parliamentary Secretary to Lord Privy Seal, and Minister of Aircraft Production, 1942–45; Parliamentary Secretary to Ministry of Transport 1945–47; Minister of Supply 1947–51.

BIBLIOGRAPHY

MANUSCRIPT SOURCES

Bank of England Archives (Bank of England, London)
Baring Brothers Archives (Baring Brothers & Co., London)
Board of Trade Papers (Public Record Office, London)
British Iron and Steel Federation Papers (British Steel Corporation Record Centre, Irthlingborough)
R.A. Butler Papers (Trinity College, Cambridge)
Cabinet Papers (Public Record Office, London)
Lord Chandos Papers (Churchill College, Cambridge)
Conservative Party Archives (Bodleian Library, Oxford)
H.F.C. Crookshank Papers (Bodleian Library, Oxford)
Walter Monckton Papers (Bodleian Library, Oxford)
Morgan Grenfell Papers (Morgan Grenfell Group, London)
Lord Swinton Papers (Churchill College, Cambridge)
Treasury Papers (Public Record Office, London)
Lord Woolton Papers (Bodleian Library, Oxford)

PRINTED PRIMARY SOURCE

Hansard, *Parliamentary Debates*, 5th Series

BOOKS

Burk, Kathleen, *Morgan Grenfell 1838–1988: The Biography of a Merchant Bank* (Oxford: Oxford University Press, forthcoming).

Burn, Duncan, *The Steel Industry 1939–1959* (Cambridge: Cambridge University Press, 1961), 728 pp.

Burn, Duncan, ed., *The Structure of British Industry, Vol.1* (Cambridge: Cambridge University Press, 1958), 403 pp.

Butler, David & Gareth Butler, *British Political Facts 1900-1985* (London: Macmillan Press, 1986), 536 pp.

Buxton, Neil K. and Derek H. Aldcroft, eds., *British Industry Between the Wars: Instability and Industrial Development 1919–1939* (London: Scolar Press, 1979 [1982 pb edit]), 308 pp.

Donoughue, Bernard and G.W. Jones, *Herbert Morrison: Portrait of a Politician* (London: Weidenfeld and Nicolson, 1973), 696 pp.

Foot, Michael, *Aneurin Bevan 1945–1960* (London: Davis-Poynter, 1973), 692 pp.

Gilbert, Martin, *Winston Churchill. Volume VIII: 'Never Despair', 1945–1965* (London: Heinemann, 1988), 1438 pp.

Harris, Kenneth, *Attlee* (London: Weidenfeld and Nicolson, 1982), 630 pp.

Jeremy, David J., ed., *Dictionary of Business Biography*, 5 vols (London: Butterworths, 1985).

Keeling, B.S and A.E.N. Wright, *The Development of the Modern British Steel Industry* (London: Longmans, 1964), 210 pp.
Kent, Harold S., *In on the Act: Memoirs of a Lawmaker* (London: Macmillan, 1979), 273 pp.
Morgan, Janet, ed., *The Backbench Diaries of Richard Crossman* (London: Hamish Hamilton and Jonathan Cape, 1981), 1136 pp.
Morgan, Kenneth O., *Labour in Power 1945–1951* (Oxford: Oxford University Press, 1984), 546 pp.
Nicholas, H.G., *The British General Election of 1950* (London: Macmillan & Co. Ltd, 1950), 353 pp.
Ovenden, Keith, *The Politics of Steel* (London: The Macmillan Press Ltd, 1978), 261 pp.
Pimlott, Ben, ed., *The Political Diary of Hugh Dalton 1918–40, 1945–60* (London: Jonathan Cape, 1986), 737 pp.
Sayers, R.S., *The Bank of England 1891–1944* (Cambridge: Cambridge University Press, 1976 [1986 pb edit]), 680 pp.
Seldon, Anthony, *Churchill's Indian Summer: The Conservative Government 1951–55* (London: Hodder and Stoughton, 1981), 667 pp.
Taylor, Eric, *The House of Commons at Work* (London: The Macmillan Press Ltd, 1979, 9th edit), 190 pp.
Teichova, Alice, *An Economic Background to Munich* (Cambridge: Cambridge University Press, 1974), 422 pp.
Turner, John, ed., *Businessmen and Politics: Studies of Business Activity in British Politics 1900–1945* (London: Heinemann Educational Books, 1984), 200 pp.
Wigg, Lord, *George Wigg* (London: Michael Joseph, 1972), 384 pp.
Wilson, Charles H., *A Man and His Times: A Memoir of Sir Ellis Hunter* (Newman Neame, 1962), 53 pp.

ARTICLE

Roberts, Richard, 'Election Values', *Financial Times*, 30 May 1987.

NEWSPAPERS AND PERIODICALS

The Daily Telegraph
The Economist
The Financial Times
The Scotsman
The Times

INDEX

Iron and Steel Board (1953–67), 6–7, 26, 32, 36, 43–4, 49–50, 52–4, 58–9, 62–4, 75–9, 81–8, 90, 92, 108, 110–11, 139, 146–7
Iron and Steel Control Board (1946–49), 19, 22–3, 26, 31–2, 43–4, 77